Understanding Bioinformatics

Understanding Bioinformatics

Julian Simpson

MURPHY & MOORE
www.murphy-moorepublishing.com

Murphy & Moore Publishing,
1 Rockefeller Plaza,
New York City, NY 10020, USA

Visit us on the World Wide Web at:
www.murphy-moorepublishing.com

ISBN: 978-1-63987-545-0 (Hardback)

Cataloging-in-Publication Data

Understanding bioinformatics / Julian Simpson.
p. cm.
Includes bibliographical references and index.
ISBN 978-1-63987-545-0
1. Bioinformatics. 2. Computational biology. 3. Biology--Data processing.
4. Systems biology. I. Simpson, Julian.
QH324.2 .U53 2022
570.285--dc23

Table of Contents

Preface

Bioinformatics is a hybrid science that connects biological data with techniques for information storage, distribution, and analysis to support multiple areas of scientific research including biomedicine. Bioinformatics is dependent upon high-throughput data-generating experiments that consist of genomic sequence determinations and measurements of gene expression patterns. As an interdisciplinary science domain, bioinformatics utilizes and amalgamates biology, mathematics and statistics to analyze and interpret biological data. It encompasses biological studies that utilize computer programming as a part of its methodology as well as a specific analysis pipeline that is used in the field of genomics. In the field of medicine especially, a number of important applications for bioinformatics have been discovered. This book provides comprehensive insights into the field of bioinformatics. It picks up individual branches and explains their need and contribution in the context of a growing economy. As this field is emerging at a rapid pace, the contents of this book will help the readers understand the modern concepts and applications of this subject.

A short introduction to every chapter is written below to provide an overview of the content of the book:

Chapter 1 - Bioinformatics is an interdisciplinary field which combines mathematics, information science and biology for understanding biological data. Structural bioinformatics, integrative bioinformatics, translational bioinformatics, animal bioinformatics, plant bioinformatics, etc. fall under its domain. This chapter serves as an introduction to the different aspects of bioinformatics; Chapter 2 - Sequence analysis refers to the way in which proteins or DNA sequences are subjected to a wide range of analytical methods to understand its features, functions, structure or evolution. This chapter deals with sequence analysis and its various aspects like DNA sequencing, sequence assembly, sequence alignment, multiple sequence alignment, etc. in an in-depth manner for the benefit of the reader; Chapter 3 - The identification and comparison of genomic features such as DNA sequence, gene expression and structural variation is known as genome analysis. It falls under a branch of biology known as genomics. This chapter has been written carefully in order to provide a better understanding of all the topics related to genome analysis ; Chapter 4 - An algorithm is a set of rules or procedures to be followed by a computer for solving problems. Bioinformatics makes use of algorithms which have been adapted for application in solving problems related to molecular and genome biology. Baum-Welch algorithm, Needleman-Wunsch algorithm, Hirschberg's algorithm, etc. are a few examples of bioinformatics algorithms which are discussed thoroughly in this chapter; Chapter 5 - Bioanalytical techniques can be defined as the various methods used for analyzing biotics and biological matrices of biological molecules. Computational biology is one such method which employs algorithms and models to interpret biological data. Bioanalysis and bioassays are other aspects which fall under this domain. This chapter aims to provide a comprehensive understanding of all the different bioanalytical methods and techniques.

I extend my sincere thanks to the publisher for considering me worthy of this task. Finally, I thank my family for being a source of support and help.

Julian Simpson

Chapter 1

An Introduction to Bioinformatics

Bioinformatics is an interdisciplinary field which combines mathematics, information science and biology for understanding biological data. Structural bioinformatics, integrative bioinformatics, translational bioinformatics, animal bioinformatics, plant bioinformatics, etc. fall under its domain. This chapter serves as an introduction to the different aspects of bioinformatics.

The Bioinformatics began with efforts to combine experimental data with computational algorithms for generation of information and knowledge about biological systems. The ultimate goal is to model complete biological processes. Biological processes are recognized series of molecular functions, with a defined starting and an ending function, relevant to integrated functioning of cells. Bioinformatics began on a small scale with sequence alignment algorithms for searching similar sequences conserved during evolution. Now, Bioinformatics encompasses new foundations for collection, organization and mining gene/protein sequences, three dimensional structures and biochemical functions, for modeling biological processes of functioning cells.

Classical Example Revealing the Power of Bioinformatics

Classical example of bioinformatics is provided by computational gene hunting for the fatal disease, cystic fibrosis (CF). This disease is associated with recurrent respiratory infections and abnormal secretions. CF is diagnosed in one out of every twenty five hundred persons in Caucasian race. One Caucasian out of 25 is a carrier of the faulty gene. Its gene is autosomal and recessive. Autosomal means a gene on chromosomes other than X- and Y-chromosomes of humans i.e. rest of the 22 pairs of chromosomes and recessive means when both alleles on each of the homologous chromosome are same. Now the presence of CF disease in other countries such as in India is well documented with the advancement of laboratory testing. Most of the phenotype of CF disease was in accordance of western population.

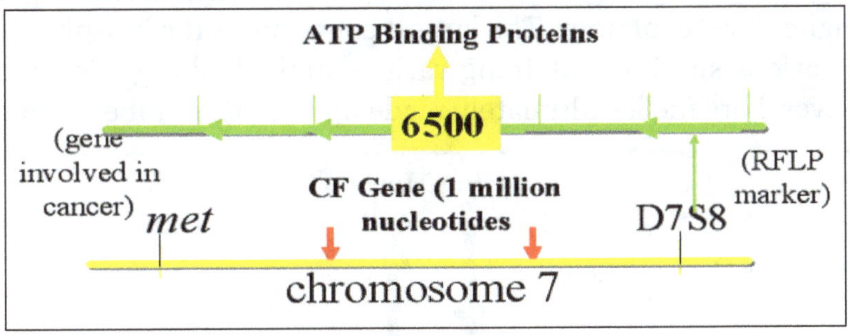

In 1980s, biologist knew nothing about this gene. The search of the CF gene began in early 1980s with genetic mapping using recombination frequency with genes and RFLP probes as markers. In 1985, three groups of scientists independently announced the gene (approximately 1 million nucleotides) on chromosome number 7, located between met gene (a gene involved in cancer) and D7S8 (an RFLP marker).

This was followed by physical mapping using chromosome mapping and chromosome jumping techniques to find actual nucleotide distance and base sequence. A clone hybridizing with D7S8, RFLP marker was selected and sequenced followed by selection of next clones hybridizing with ends of the clone. After several iterations of selection and sequencing, the gene of interest will also get sequenced, if it is located near to the starting marker. Knowing the approximate location of the gene, does not lead to the gene itself. For example, the Huntington's disease gene was mapped in 1983, i.e. before CF gene and remained elusive till 1993. In addition, after sequencing we don't have any idea about the function of the gene of interest. However, with database search in 1989, a region encompassing about 6500 nucleotides was found to have sequence similarity to ATP binding proteins such as Cystic Fibrosis Trans membrane Conductance Regulator (CFTR). The CFTR is an integral cell membrane single polypeptide chain of 1480 amino acids with five domains. The domains include two trans membrane, two cytoplasmic and one regulatory. This shows that each domain has a specific function involving transport of Chloride ions through the cell membrane. Consequently, mutations in these domains will affect transport of Chloride ions through cell membrane causing abnormal secretions to manifest CF disease. This sequence database search, based on sequence alignment, not only identified the gene but also hinted at the possible structure and its function, through which the disease was manifesting. This showed the power of sequence alignment for searching similar sequences in the database. This sequence similarity database search is most widely used tool of bioinformatics now-a- days to have an idea about the function of a given protein sequence. This was the first example of revealing a protein structure to suggest biochemical function and then phenotype. This laid the foundation of Bioinformatics.

Biochemistry and Bioinformatics

There are two types of systems or machines: living and nonliving. Both types of systems or machines transform matter, energy and information. Biochemistry is the study of chemistry of life. Chemistry is the making and breaking of chemical bonds. Life is defined as the capability of a system to reproduce self-i.e. to produce a replica of the system by itself. However, only living things can reproduce self and this ability is made possible through making and breaking of the chemical bonds at very rates. In addition, the living cells/machines are capable of self-duplication with modifications i.e. evolution through survival of the fittest. Living systems or machines are hierarchically organized in the form of an individual component in a container, where a container may be a component in a higher level container. The largest container is the biosphere containing various ecosystems having various species containing various individual organisms made up of various types of cells with several organelles ultimately made up of large number of macromolecules.

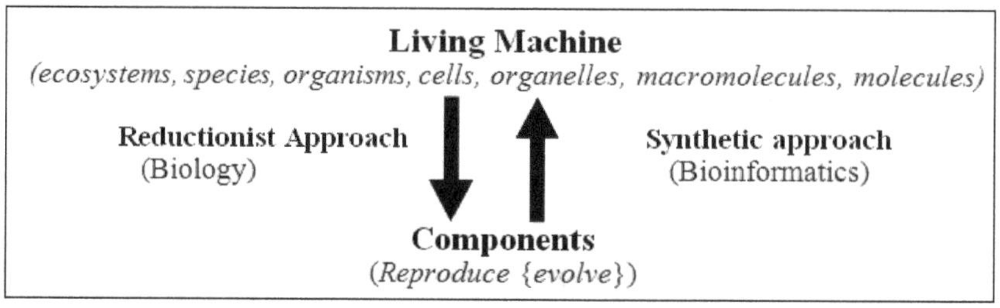

Biochemistry and Molecular Biology is concerned about collection and interpretation of data about living components, under normal and abnormal situations, so as to take corrective measures, if any of the living components behaves in some undesirable manner. The approach followed has been reductionist with isolation of a living component and then making observations. With the advent of modern sophisticated high-throughput technologies in biology, the quantity of data produced is so large that it is impossible to manage this large scale data without automation using computing and robotic devices. Consequently, mathematics, computer science and information technology were increasingly applied to collect, organize and interpret this large quantity of data acquired through automated high throughput technologies, which are low cost and very fast. Therefore, high-throughput technologies are those technologies which are able to acquire and analyses large quantities of nucleic acid sequences (DNA/RNA), protein expression in a cell, drug activities etc., at a very high speed. These high-throughput technologies include next generation sequencing for determination of sequences of complete chromosomes/genomes, an area called genomics; microarrary technology to study RNA expression of complete genome under a given set of conditions, an area called transcriptomics; 2D polyacrylamide gel electrophoresis to study expression of complete proteins under a given set of conditions, an area called proteomics; screening of a large number of drug molecules, an area called pharmacogenomics and large scale modelling of biological information, an area called systems biology. Bioinformatics has emerged as a result of the advancement of the experimental technologies for collection of these vast amounts of data in biochemistry and molecular biology and subsequent organization, interpretation and modelling these data with mathematics using computers. Bioinformatics, therefore, fuses biology with mathematics and computer science. Bioinformatics follows synthetic approach i.e. it aims to assemble the small molecules such as amino acids and nucleotides into proteins and nucleic acids. Then the macromolecules such as proteins, nucleic acids, polysaccharides, lipids etc. into organelles such as ribosomes, cell membrane, chromosomes etc. These organelles then are aimed to be assembled as organelles, then cells and then individuals.

This is transformation from basic reductionist approach of observing single component to a synthetic approach to organize individual components on large scale for developing models capable of making predictions about living systems. In Biochemistry we are concerned up to macromolecules and this field is known as Molecular Bioinformatics. Here, we will be concerned about the Molecular Bioinformatics i.e. collections of molecules: genes, proteins, lipids, carbohydrates etc. and their interactions.

Central Paradigm of Bioinformatics

We are well aware that biology at the molecular level is guided by central dogma of molecular biology. However, this dogma is not able to answer questions such as when a gene in the life cycle or in which tissue or why and how a gene/protein is expressed in an individual. Central dogma of molecular biology tells only the flow of information from DNA to RNA (transcription) to Protein (translation) but never back from protein to RNA or from protein to DNA. However, information from RNA may be converted back to DNA as in reverse transcription. Central dogma of molecular biology simply conveys that sequence information determines phenotype of a living system.

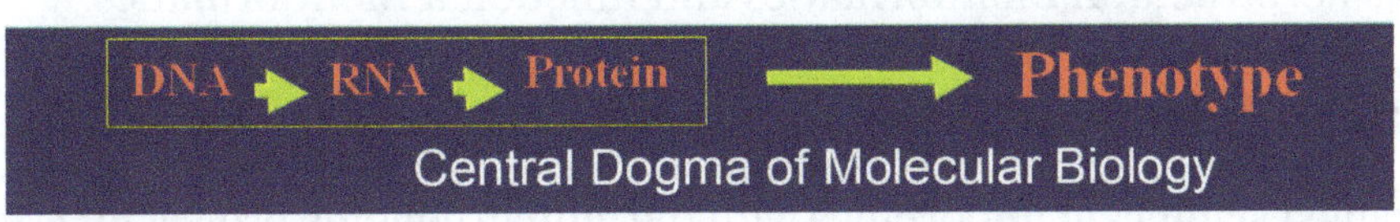

Therefore, central dogma of Molecular biology dictates the direction of synthesis of biological macromolecules i.e. DNA to RNA and then translation to proteins. This dogma forbids the syn-thesis of nucleic acids from proteins. This flow of information then decides the manifestation of phenotype of a given cell. However, the central dogma of molecular biology does not specify the way in which the phenotype of a cell will be achieved.

On the other hand, the central paradigm of bioinformatics considers that sequence information in DNA, RNA and proteins represents same knowledge and determines the molecular structure of actual transforming machines i.e. proteins and RNAs, to interconvert/ transform metabolites and therefore controls the phenotype of a biological system. Therefore, central paradigm of Bioinformatics specifies that the sequence information (DNA/RNA/Protein) determines the protein structures, which in turn will give specific biochemical functions, to finally express the cellular phenotype through interaction of metabolic, signaling and gene expression pathways. The systems/ cells which will be surviving will then reproduce to pass on the genetic sequence information (DNA/RNA/Protein) to next generation cells, which will repeat the above life cycle. Therefore, Bioinformatics is having four sequential stages. The first is genetic sequence information determination encompassing genes and proteins. The next is the transformation of this sequence information into three dimensional structural information of proteins. The classical experiment of Christopher Anfinsen in 1973 using RNAse denaturation and re-naturation established that the information for folding a protein into three dimensional structures is encoded within the amino acid sequence of the protein itself. Therefore, once the protein is synthesized, it will fold into the three dimensional structure as per information contained in its sequence. Once the protein folds into a defined three dimensional structure, then this protein and other structures are actual workers of the biological processes, which brings about the biochemical transformations using pathways. These pathways then interact with each other to express the phenotype of the cell. Those phenotypes which survive this life cycle are able to reproduce the next generation.

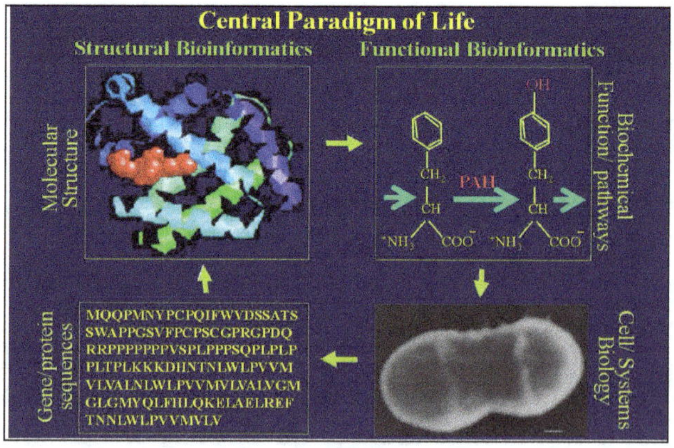

Bioinformatics encompasses computational molecular biology tools, i.e. mathematical and computational analysis for sequences and structures of biological macromolecules, in large scale genome research, targeting investigation of the complete set of chromosomes & genes of an organism and their RNA as well as protein products. On the basis of the central paradigm, bioinformatics has two key areas i.e. structural Bioinformatics and functional Bioinformatics.

The Goal, Aim and Scope of Bioinformatics

Bioinformatics gained significant importance with the human genome project for creation of detailed genetic and physical maps, and DNA sequence for each of the 24 different human chromosomes. It encompasses development of automated technologies for generation, organization, mining and modelling of biological data with the goal of revealing new insights, theories and principles about living things, so as to enable predictions about them. Most of the organized activities of bioinformatics has focused on the collection, organization, interpretation and modelling these processes on a large scale consequently giving rise to sub-areas such as genomics, transcriptomics, proteomics and metabolomics. And all this for better nutrition, health and environment of living machines through predictions from modelled systems.

If we are interested to take a snapshot from history of life that extends back some 2.5 billion years, then we find no biological experimental way except to have some information from DNA of fossils. On the other hand, with bioinformatics, we can expect to model functional systems in that age. Similarly, we can also expect to predict that in future there may be evolution of a deadly virus and therefore design strategies in advance, to combat evolving deadly virus. Imagine a scenario in future in which a new biological virus creates an epidemic of some fatal disease. Laboratory biologists will provide the genetic material for automated sequencing and computer programs will then take over to design an effective drug against this virus. Comparative genomics of this new virus with previously studied viral genomes will suggest some targets for antiviral drugs. Three dimensional structure prediction of the target shall provide the actual site of action for the drug. High throughput virtual screening may provide leads for faster drug development, which may be finally tested on bench for its efficacy. Therefore, the scope of bioinformatics is to model living systems so to predict the behaviour of living systems under perturbed conditions.

The Emergence of the Discipline of Bioinformatics

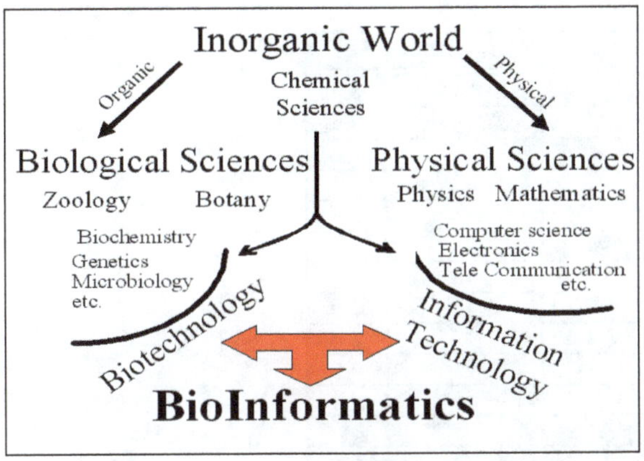

Bioinformatics: Marriage of life and non-life.

Bioinformatics is a young but fast-growing field for biological data collection, organization, interpretation and modelling. Tools and techniques for bioinformatics are derived from multidisciplinary combinations of varied disciplines from natural and physical sciences. Previously various disciplines were carved out as and when sufficient specialization was achieved. However, now bioinformatics is borne out of alliance between existing disciplines from life and non-life.

Structural Bioinformatics

Structural Bioinformatics refers to gene mapping and determination of DNA/RNA/protein sequences/structures on large scale. Genetic linkage maps databases store distances between genes and other markers, based on meiotic recombination frequencies. Chromosome physical maps represent distances between these genetic markers based on nucleotide lengths between them. Sequence maps store the actual nucleotide and amino acid sequences of the genes and proteins. At protein level, structural Bioinformatics also elucidates the representative protein folds.

Proteins are not only of fundamental biological importance, they also play an important role in many diseases. Because the function of a protein is linked to its structure, the structural analysis is of great importance for understanding pathological processes in order to provide new ways in diagnostic and therapeutic medicine. Therefore structural bioinformatics has a great impact to biomedical science and drug discovery. The influence of structural bioinformatics to clinical research is highly dependent on the accessibility of genomic, proteomic and clinical data and the development of methods for connecting this information with structural data. The visualization of the structural data is of crucial importance. Procedures such as: inhibition and neutralization of enzymes; interaction of a hormone with its receptor and antigen-antibody interactions cannot be adequately explained without the optical advantage obtained with such visualization methods. The structural analysis is complemented by quantum theoretical methods providing details at the electronic level, for example during the binding of a substrate to an enzyme, that remain unrevealed after structural elucidation.

Protein Biosyntheses

Amino acids	Quantity	%
Ala (A)	51	8.7
Arg (R)	24	4.1
Asn (N)	21	3.6
Asp (D)	32	5.5
Cys (C)	15	2.6
Gln (Q)	22	3.8
Glu (E)	34	5.8
Gly (G)	46	7.9
His (H)	15	2.6
Ile (I)	27	4.6
Leu (L)	55	9.4
Lys (K)	38	6.5
Met (M)	25	4.3
Phe (F)	26	4.4
Pro (P)	27	4.6
Ser (S)	34	5.8
Thr (T)	28	4.8
Trp (W)	12	2.1
Tyr (Y)	21	3.6
Val (V)	32	5.5

$$\begin{array}{c} R \\ | \\ H_2N - C_\alpha - COOH \\ | \\ H \end{array}$$

Proteins are built up as polymers of up to 20 different amino acids. The table shows the percentage of the different amino acid types constituting the enzyme glutamic acid decarboxylase. Each amino acid consists of an amino group, a carboxyl group and a variable side chain (R), which determines the type of amino acid.

Proteins are used by the cell to read and translate the genomic information into other proteins for performing and controlling cellular processes such as: metabolism (decomposition and biosynthesis of molecules); physiological signaling; energy storage and conversion; formation of cellular structures etc. Proteins are synthesized by means of the genetic code which is inherent through replication.

By considering the chemical composition of a protein, one obtains the, for organic molecules, typical chemical elements in different quantities as illustrated in the case of glutamic acid decarboxylase (GAD):

$$C_{2925}H_{4532}N_{780}O_{844}S_{40}$$

However, the chemical composition delivers little information about proteins. A better understanding of the protein molecules is obtained by considering their modular composition. Proteins contain amino acids as subunits. Amino acids consist of an amino group, a carboxyl group and a variable side chain. The proteins are built up as polymers of the amino acids. The amino acids are connected by a peptide bond between the basic amino group and the acidic carboxyl group of each amino acid in the polypeptide chain. 20 different amino acids are involved as residues in protein sequences. Each of the 20 amino acids differs by its side chain, which is responsible for the physicochemical properties such as: electric charge; polar, non-polar; acidic; basic; hydrophobic etc. Some of the side chains have ring like structures (aromatic), while others consist of straight carbon chains (aliphatic). The different amino acids are represented in different quantities in a protein. Hypothetically, were one to mix the amino acids in the right quantities in a test tube, one would never obtain the desired protein.

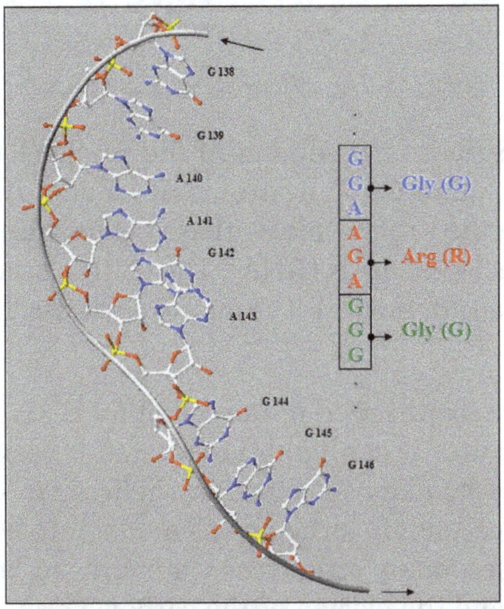

Messenger RNA (mRNA) contains the information needed for protein biosyntheses. The information is determined by a sequence of 4 nucleotides: adenine (A), cytosine (C), guanine (G) and uracil (U). An amino acid is encoded by a triplet of nucleotides. This encoding is redundant in that sense, that one amino acid can be encoded by different triplets.

The protein and its structure are not only determined by the percentage of the different amino acids but also the information about the sequence of the amino acids in the polypeptide chain is necessary. The information encoded in the amino acid sequence determines the 3D structure of a protein and in consequence its function. The protein biosyntheses is determined by the genetic code in the DNA (deoxyribonucleotid acid) through the intermediate RNA (ribonucleotid acid). During gene expression the DNA sequence of the gene is copied into messenger RNA (mRNA). The mRNA then serves as template for the protein biosyntheses. It contains a sequence of 4 nucleotides: adenine, cytosine, guanine and uracil, which determine the copied genetic code. To encode an amino acid, 3 nucleotides are combined in a triplet (codon). Because 64 possible combinations of the 4 nucleotides in triplets are encoding 20 different amino acids the code is redundant. The ribosome is the biological unit in the cell where the copy of the genetic code in the mRNA is translated into the amino acid sequence and the protein is build up. The ribosome is a complex of different ribosomal proteins and RNA. During the translation the condons of the mRNA are recognized by the complementary nucleotide triplets (anticondons) of the transfer RNA (tRNA) and a condonanticondon bonding results. Every tRNA molecule transports an amino acid, where the type of the amino acid is dependent from the anticodon. During the interaction of the mRNA and the tRNA the amino acids are connected according to the encoded information in the mRNA sequence.

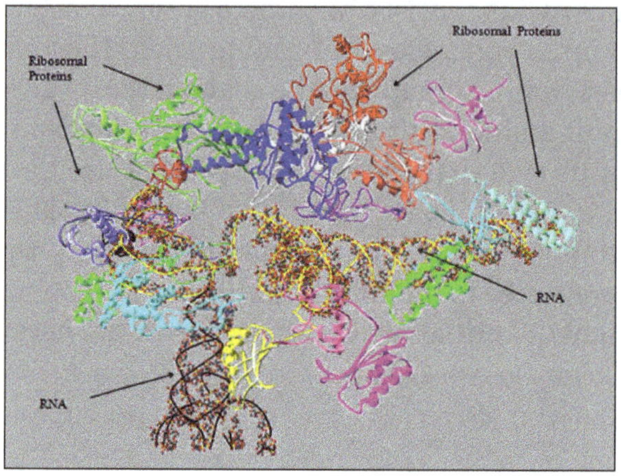

Figure shows parts of the 30S ribosomal subunit of the whole ribosomal complex with transfer RNA (tRNA) and mRNA (resolution 6.46 angstroms, R-value = 0.354). The ribosome is the biological unit in the cell where the protein biosyntheses takes place. It is a complex of different ribosomal proteins and ribonucleotid acids (RNA). The mRNA serves as template for the protein syntheses. Each tRNA bounds an amino acid. The amino acids are then connected according to the mRNA sequence.

Once an amino acid sequence is synthesized in the cell, it folds together to a well defined and, for its sequence, unique 3D structure. It can be differentiated between the primary structure (the sequence of the residues), the secondary structure (α -helices, β -sheets and loops) and the tertiary structure (folding of the secondary structure elements into a three dimensional structure). The polypeptide chain forms secondary structures via hydrogen bonds (interaction between the hydrogen of the amino group and the oxygen of the carboxyl group) between amino acids in the chain. The secondary structure elements are then folded, through further interactions between the residues, into the three dimensional structure.

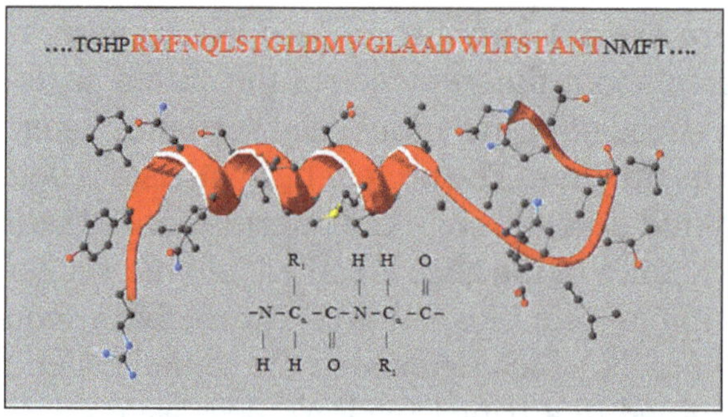

The 3D structure of a protein is uniquely determined by its amino acid sequence. In the polypeptide chain the amino acids are connected via the respective carboxyl group and amino group.

The main driving forces in protein folding are hydrophobic and electrostatic interactions. Domains are compact regions of structure within the larger protein structure. Every domain has its own hydrophobic core and is connected to the rest of the protein via the polypeptide backbone. Normally a domain is collinear with a defined sequence part and the protein structure is divided into several domains. Many proteins are oligomeric, which means they are composed of more independent polypeptide chains (quaternary structure: three dimensional arrangements of the monomers). The individual polypeptide chains in the quaternary structure can interact with electrostatic forces and hydrogen bonds.

Functional specificity of a protein is linked to its structure. Due to the folding structure each of the critical residues responsible for the protein function is brought into a precise geometric arrangement. They are building the active site: a localized combination of amino acids within the tertiary structure that acts with other molecules and provides the protein with biological activity. Then interactions of a protein with other molecules are determined by residues which are close in 3D space but may be very distant in the amino acid sequence. The active site is then often found only in a small part of the structure and the rest of the protein structure is mainly necessary to enable and maintain the correct spatial position between the amino acids on the active site. Therefore, to understand a protein function, the 3D structure of the protein reveals far more information, than its sequence.

The knowledge of the 3D structure is essential to understand the function of proteins and the interactions between proteins within the organism resulting in metabolism, reproduction and form. Classes of proteins are: enzymes; hormones; regulators; signal receptors; ion channels; antibodies etc.

Structural Classification of Proteins

The evolution of proteins is restricted due to physical chemical reasons. Solvent accessibility and hydrophobicity are playing an important role in protein folding. Proteins always exist in aqueous solution and they constantly interact with water molecules. The side chains of the amino acids are divided into two groups: the hydrophilic and the hydrophobic side chains. Water molecules have polar charges and liquid water consists of an uninterrupted lattice of hydrogen bonded molecules. A non-polar protein molecule dissolved in water interrupts the regular hydrogen bond lattice, which is energetically unfavorable because the water molecules are forced to work around the globular protein and they form a kind of cage. Therefore to diminish the energy costs of solvating the protein, it is favorable to interact with the water molecules and the amino acids with hydrophilic properties have to be arranged solvent exposed at the protein surface, whereas the hydrophobic amino acids are mainly concentrated in the protein core. The number of possible occurring structure motifs (folds) is therefore limited, due to these physical reasons, and they are repetitively used in different proteins. This physical aspect is used for a hierarchical classification of protein structures. Proteins are grouped according to what kind of secondary structure elements they have. Within those larger classes, subclasses are based on the kind of arrangement of the secondary structure elements in the protein. The focus in protein classification is the arrangement of proteins that have similar structural architectures; it doesn't matter if their amino acid sequences are related. The same physical reasons, leading to the limited number of different occurring folds in proteins and their hierarchical classification, are the basis for the success of many algorithms in structural bioinformatics such as: the structural alignments for the search of homologue proteins and the homology modelling.

Protein Structure Database

Protein structural information is publicly available at the protein database (PDB), an international repository for 3D structure files. PDB is the most important source of protein structures. At the moment PDB contains more than 52.000 protein structures. Before a new structure of a protein is inserted into the database, a time consuming and careful examination of the data must be carried out to guarantee the quality of the structure. Access to the database is enabled via the Internet.

Access to the Structural Data

The entry point to the structural protein data is the PDB. The search for a particular protein structure of interest can be initiated by entering the 4 letter PDB identification code at the PDB main page. Alternatively, the PDB can be searched by use of keywords. A comfortable access to the PDB database is enabled by the integrated ENTREZ search and retrieval system of the NCBI (National Centre for Biotechnology Information). This can be done by searching the structural database using specific keywords such as: the name of the protein or organism or other identifying features such as: domains; folds; structures of the protein bound to a substrate etc. The keywords are used to search for entries in the most important fields in the PDB data header. The advantage of the access via ENTREZ is the availability of several public domain tools like BLAST (Basic Local Alignment Search Tool) which enables the user to compare an uncharacterized protein sequence with the whole sequence database or VAST (Vector Analysis Search Tool), which is useful for the determination of structural neighbors of a given protein structure.

After the successful access to the protein of interest in the database, the appropriate information is available through the Information Portal to Biological Macromolecular Structures interface which offers, among others, the opportunity to:

- View protein structures with simple 3D display tools.
- Display PDB file and Header in HTML format and download structure files.
- Display geometrical properties, such as: dihedral angles; bond lengths and bond angles, in tabular format.
- Display and download sequences in FASTA format.

A large number of protein structures are deposited in the PDB structure database. The structural information is stored in the PDB data files as atomic coordinates. Information about protein structures is available through the Information Portal to Biological Macromolecular Structures interface. In the database, the proteins are identified by a 4 letter code. The available information includes: sequence of the protein; experimental parameters; structural properties such as dihedral angles; bond angles; bond lengths and others.

Additionally, links for many tool databases for protein structure analysis and protein classification databases (similar to SCOP) are offered. The retrieved protein structure is available as a structure file which contains the atomic coordinates of the protein crystal structure.

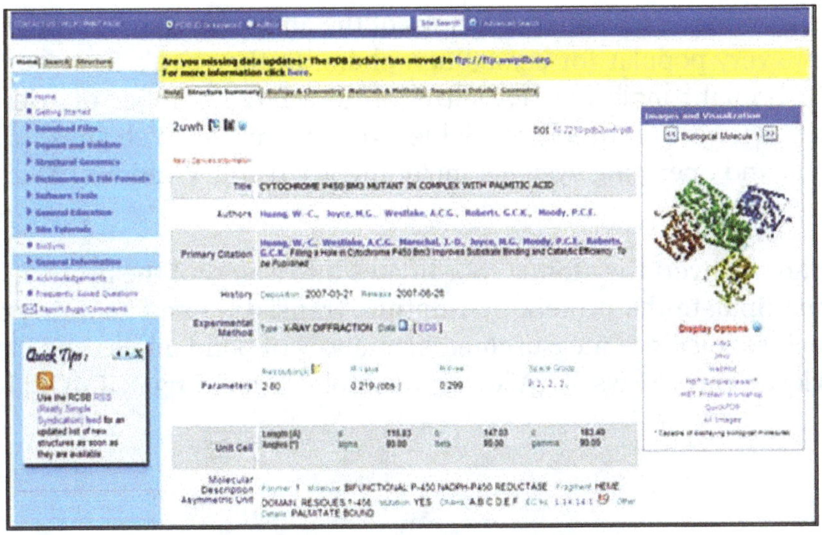

Data Format

The Brookhaven PDB format is the most commonly used protein structure data format. This format is divided into a section containing miscellaneous information and a section containing the atom records. The main parts of the PDB data format are:

- HEADER: contains name and source of the protein, resolution, description of experimental conditions and details, names of the authors, crystallographic parameters including R-factor, sequence information, secondary structure information, literature citations etc.
- SEQRES: this part contains the protein sequence.
- ATOM: this part contains the atomic coordinates and the B-factor values as part of the protein chain.
- HETATM: coordinates of cofactor molecules, substrates, and other groups that are not covalently bound to the protein.

PDB files are used as input for protein visualization and they offer the necessary information for the calculation of protein properties, manipulation and representation of the structures. Other commonly used formats are: mmCIF (macromolecular Crystallographic Information Format) and the ASN.1 format.

From the PDB database the structure files can be downloaded and the protein visualized at the local place.

Visualization of Protein Structures

The first computer representations of macromolecular structures were made during the 1960's. Researchers at the MIT developed a system based on an oscilloscope for displaying rotating wire-frame representations of macromolecular structures. At the same time, a program to produce stereoscopic drawings of molecular structures with a pen plotter was developed. In the mid 1970's a protein structure was visualized entirely with computers for the first time. During the 1980's computer systems showing wireframe renderings of the amino acid chain, which could be rotated in real time become very popular for crystallographers. In the 1990's first programs running on Macintosh and PC brought molecular visualization to a large number of scientists. Nowadays, a number of protein viewers are freely available from referenced Web sites, which run on most computer platforms and operating systems including Microsoft Windows, Macintosh and UNIX X-Windows.

Such viewer programs convert the atomic coordinates into a view of the protein. Common viewers provide ways to manipulate the protein by rotating, translating and zooming the molecule and enable features such as: distance measurements; analysis of bond angles and torsions angles etc. Commonly used viewers are: Swiss-PdbViewer; RasMol; Chime; Cn3D; JMOL; KiNG etc.

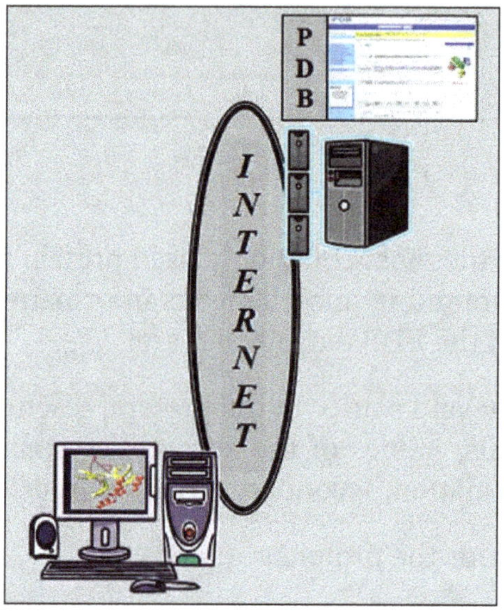

The PDB database can be accessed through the Internet. The PDB data files are the input files commonly used, for protein visualization. The data files are downloaded from the PDB database, and the protein visualized, locally.

Protein Models

In the world of protein molecules the application of visualization techniques is particularly rewarding, for it allows us to depict phenomena that cannot be seen by any other means. Being able to "see" the 3D structure of a protein and to analyze its shape is of crucial importance for understanding protein properties and interactions. Obviously, it is not possible to see a protein for example by a microscope with X-rays focussing lenses. Therefore there exists no real image, like a microscopic view from cell, of a protein. Instead a model, based on atomic coordinates resulting from the optimal fitting into the electron density map, must be used. (To be precise: the real experimental results in X-ray crystallography are the diffraction patterns on the detector, the electron density map is already derived from these data). This means that there is always an element of interpretation of the experimental data that leads to a model - a hypothesis about the structure that gave rise to the experimental data that we collected. This model of a structure is a three-dimensional representation of a protein that contains information about the spatial arrangements of groups of atoms. The model reflects the experimental data in a consistent way. With the aid of the model, experimental results concerning structural properties can be interpreted and further hypothesis about the protein interactions be formulated.

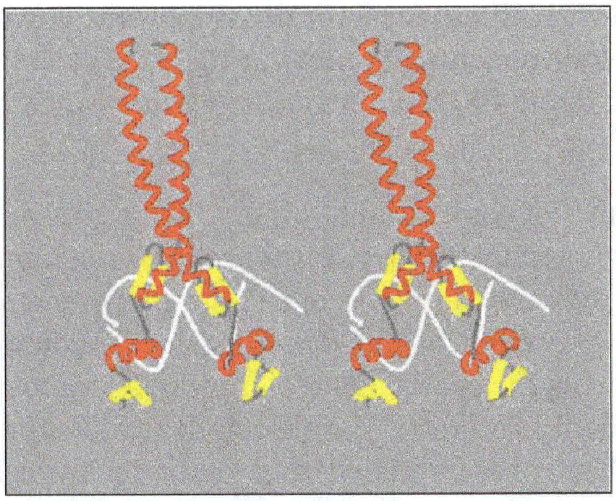

Figure above show Stereo view of a zinc finger domain (involved in the specific modulation of gene expression) interacting with a DNA double strand (resolution: 1.5 angstrom, R-value = 0.216). The two images of the protein are rotated around the vertical axis by 2 degrees. The principle of seeing a 3 dimensional image is based on the fact that each eye sees a slightly different side of the protein and the 2 views are combined mentally to a 3D object. By fixing the left image with the left eye and the right image with the right eye a 3D impression results (good luck).

However, simply plotting the coordinates is not suitable for visualization, for this, the connectivity between atoms has to be taken into account. This can be done by the "chemistry rule method". This method uses some of the rules of chemical bond lengths to connect atoms. For example, all neighboring carbon atoms at a distance of 1.5 angstrom will be connected by a chemical bond. If such rigid chemistry rules are implemented in the viewing program, then no complete bonding information is necessary in the structure file. This approach is based on the PDB files. Another method to connect atoms in a protein is the "chemical graph method". By this method residue dictionaries, containing a list of the amino acids with their respective standard atoms and bond lengths, are used to construct a chemical graph by considering sequence information. Then the viewing program uses the information from the dictionary to connect the atoms. Structure files (MMDB: Molecular Modelling Database type) containing residue dictionaries are derived from PDB files.

Representation of Protein Structures

The information stored in the structure file is then visible as 3D graphic. The proteins are so complex, that the 3D structures are difficult to interpret visually. The first problem in the visualization and interpretation of a protein structure is the appropriate representation. The human eye can interpret 3D solids but has difficulties with topologically complex 3D data sets. The number of amino acids in proteins ranges from 50 to 2000 residues. For example: the human serum albumin protein contains 4902 atoms. Because of the complex protein structure, special simplified representations are necessary. Although the representations are constructed, they are based on experimental data and therefore they represent real aspects of proteins. Successively higher grades of abstraction of the representations are reached.

The basic representation of protein structures is the balland-stick model where atomic details are visualized. The covalent bonds are represented as sticks between atoms, which are represented as spheres or simply coloured junction points of the sticks. Non covalent bonds such as hydrogen bonds are represented by dotted lines. Structurally important covalent bonds such as sulphur bonds are highlighted by colour encoding (yellow in the case of sulphur bonds). The atoms are generally coloured according to the CPK colour scheme: white for carbon; red for oxygen; blue for nitrogen and yellow for sulphur.

Additionally there are a number of conventionally simplified representations of protein structures that allow the visualization of overall topology, without the confusion of atomic details. A common feature of all proteins is that they are composed of a linear polymer chain, the backbone. In the native state of a protein the polymer chain is folded in 3D space. The first step is the representation of the backbone as a string showing the spatial course, from N-terminus to Cterminus, of the protein backbone. This allows the identification of simple topological properties (for example, whether the protein has a spherical shape or is stretched) and facilitates the relation of the sequence of the protein to its spatial structure.

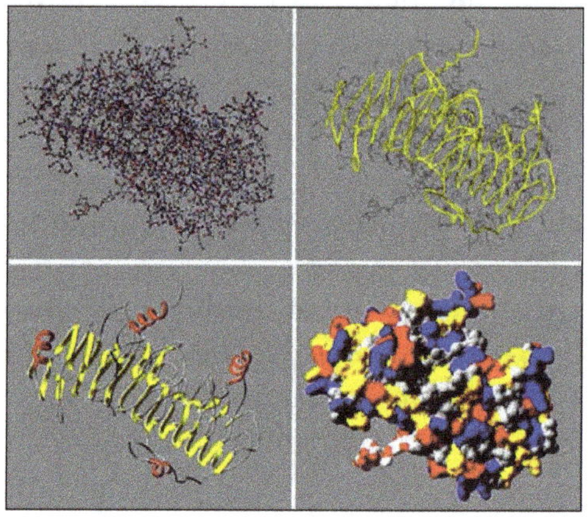

Figure shows representations of protein structures illustrated in the case of chondroitinase B (enzyme that cleaves the glycosaminoglycan, dermatan sulphate) with the beta solenoid domain (resolution: 1.7 angstrom, R-value = 0.19). Top: Left picture: Ball-and-stick representation. The atoms of the backbone and the side chains are connected according to chemistry rules. Right picture: Representation of the backbone as a string. Bottom: Left picture: Representation of the secondary structure elements as ribbons: -alpha helices (red) and - sheets (yellow) connected by loops in grey colour. Right picture: The molecular surface is coloured by chemical type. Non polar residues are gray, polar residues are yellow, basic residues are blue and acidic residues are red.

Another simplification of the representation and visualization of proteins is based on ribbons and helices, allowing the user to visualize essential features of the protein topology as secondary structure elements. Protein coordinate data sets don't come labelled with secondary structure classifiers. Secondary structures can be detected by their hydrogen bond patterns and their backbone torsion angles. Secondary structure elements can also be represented as icons (cartoon representation), whereas the -helices are represented as cylinders and the -sheets as arrows. Visualisation and analysis of the secondary elements plays, amongst other things, an important role in protein classification.

The proteins can further be represented by their molecular surface. The molecular surface is the boundary of the molecule volume within which no other molecule can enter. Molecular surfaces are helpful for the description of the topological properties of the protein shape. The surface is defined by the van der Waals radii of the atoms, which is the radius of the sphere around the atom centre, inside which another atom's spherical boundary can't pass. Then the surface depends on the atomic van der Waals radii and the coordinates of the atoms in the molecule. The protein surface is not a quantity defined by a unique property. Surfaces can be constructed as molecular or accessible surfaces. The first kind of surface is defined as the contact surface of a spherical probe (water molecule) with the protein molecule, the second kind as the locus of the centre of the probe rolled around the molecule. The default probe radius for each type of surface is 1.4 angstroms.

An important task is the visualization of non-structure based functional annotations in protein 3D structures. Bioinformatics is a science with centralized data banks, combining sequence information to functional and structural information, which are interconnected through the Internet. An adequate representation of structural properties together with the mapping of biological and physicochemical information from different databases helps to understand proteins visually showing context and connection.

Representation of Sequence-Structure Relationships

To investigate the structure and function of a protein, researchers create increasingly sophisticated computational models for transforming genetic sequencing data into comprehensive information at the protein level to better understand biological processes.

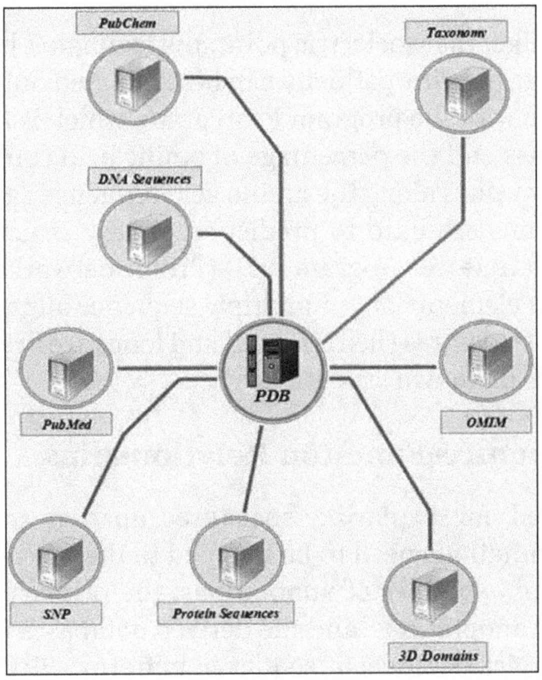

Figure above shows The PDB database is interconnected with different public databases. The compound of data bases includes molecular, biological and literature databases which are additionally interconnected to further databases. The Nucleotide database is a collection of sequences from several sources, including GenBank, RefSeq. Protein sequences result from a variety of sources, including SwissProt, PIR, PRF. PubMed is an archive of biomedical and life sciences journals. The PubChem substances database contains descriptions of chemical samples. OMIM (Online Mendelian Inheritance in Man) contains a catalogue of human genes and genetic disorders. The taxonomy database contains the names of all organisms that are represented in the genetic databases with at least one nucleotide or protein sequence. SNP (Single Nucleotide Polymorphism) contains gene variations. The 3D Domain database contains compact structural domains. The access from one data base to another is done by active links.

Special attention has been done in the development of viewers that can present structure and sequence data in a unified interface. 3D viewers and alignment editors are connected together allowing the rapid refinement of sequence-structure alignments by taking advantage of the immediate visualization of resulting insertions/deletions and strict conservations in their structural context. Multiple linked visualization of protein structure and sequence view show context maintenance by fish eye sequence view and magic lens. Additionally, web applications for the visualization of multiple protein sequences and structures, allowing the visualization of conservative and variable residues of active and binding sites in protein families were developed. Visualization of the relationship between exon distribution in the gene and protein structural elements and the association of sequence signals to 3D structure provide a valuable visualization environment for gene organization, gene evolution, protein folding and protein structure classification. The location of the exon boundaries and the intron phase are mapped onto a multiple structural alignment. Comparative analysis of exon/intron organization of genes and their resulting protein structures are important for understanding evolutionary relationships between species, rules of protein organization and protein functionality.

Physicochemical properties like: the isoelectric point; the estimated half-life of the protein; its instability index and the average of hydropathicity can be calculated out of the amino acid sequence. This can be done for example with the program ProtParam, which is additionally used for the calculation of the molecular mass and the percentage of amino acid composition of the protein. The courses of the hydropathicity values along the amino acid sequence can be determined and plotted by ProtScale. Several programs are used to predict secondary structures of proteins from their amino acid sequence. For example the program PHD (Profile network prediction HeiDelberg) predicts the secondary structure elements out of multiple sequence alignments. By use of neural networks, 3 different secondary structures (helix, strand and loop) are predicted. The neural networks are first trained with proteins of known structures.

Representation of Structure-Function Relationships

Special procedures are used for exploring structure function relationships. Therefore, sequence-based functional annotations need to be mapped to the corresponding part of the protein structure. Currently there are two types of annotations, the primary database (knowledgebase): SWISS-PROT with sequence annotations and the derived databases (cross referenced by SWISS-PROT): PROSITE with biological significant sequence patterns; PRINTS with fingerprint motifs of proteins; PFAM contains for every family a multiple sequence alignment as well as a hidden Markov model; Blocks a database for multiple alignments of motifs; Highly conserved domains in the structures can be found in the CDD: Conserved Domain Database etc.

A motif is a locally conserved region in a sequence or a short sequence pattern shared by a set of sequences. Often motifs are localized in active sites or binding sites for substrates and coenzymes. Motifs are common in protein families where the requirements of the active site mean a restriction of the protein evolution. Motifs are derived from multiple sequence alignments of members of a protein family. Common motifs in proteins often indicate a common function of proteins, even when no other sequence similarity can be found. Therefore motifs are useful for the prediction of functions of an unknown protein. PROSITE is a database which contains biologically significant patterns responsible for the function of a protein family. The PROSITE motifs consist of highly conserved residues which are relevant for:

- Catalytic sites for enzymatic function.
- Binding sites for molecules such as: DNA, ATP or other proteins.
- Binding sites for ions.

These motifs are described by the PROSITE syntax:

[FY]-x(6)-C-C-x(7)-C-[LFY]-x(6)-[LIVMFYW].

This pattern represents a common motif in the albumin family. The amino acids are represented in the one letter code x(n) means an pattern of n arbitrary residues. Letters between brackets (for example: [LFY]) mean that one of the involved amino acid must be present at the specific position. One letter (C) in the pattern means that exactly this amino acid and no other (very highly conserved) must be present at the specific position. This motif pattern is used to search and visualize the corresponding residues in the 3D structure of human serum albumin. Albumin is synthesized in the liver and serves among others as transport protein for water unsolvable substances. It binds fatty acids, hormones, bilirubin and drugs.

Essential properties of proteins, responsible for their functional behavior, can be described by use of single physicochemical parameters. This simplifies the understanding and computation of specific protein properties. For example: properties of transmembran proteins can be described by hydrophobic parameters; electrostatic interactions between an enzyme and a substrate (resulting in catalytic activities) can be described by an dielectric constant. The combination of physicochemical and structural properties of proteins may suggest: the location of catalytic sites; the location of interaction sites; identification of targets for site directed mutagenesis and the identification of membrane spanning domains. Therefore a mapping of physicochemical properties onto the protein structure is performed. To connect information about physicochemical properties with the topology, colour encoding according to the specific parameter is used.

How bioinformatics contributes to human health care will be dramatically influenced by the degree to which the tools are adopted and used as a routine aspect of biomedical research. The choice of problem based representations will improve the usability and importance of visualization tools for medical research. This will broaden the use of genomic and proteomic information in medical research and practice while enabling new insight into biological processes.

Database Aided Identification of Peptide Fragments

In order to identify a protein, it is necessary first to determine its amino acid sequence. Large proteins are decomposed into peptide fragments by an enzyme (for example: trypsin). The masses (in Dalton units) of the peptide fragments are determined by mass spectroscopy. In mass spectroscopy, the peptide fragments are ionized and then separated according to their mass using several techniques. One technique is the SELDI-TOF "Surface enhanced laser desorption ionization - time of flight" method, where the various masses of peptide fragments are determined according to their time of flight (TOF). Mass spectroscopic methods deliver, for every protein, a characteristic peptide mass fingerprint. From this fingerprint, the peptide mass spectrum, the peptide fragments can be identified by data base aided methods which allow conclusions to be drawn about the sequences of the peptide fragments and the identification of the protein. One of these tools, the program Aldente enables an automatically comparison of the determined mass spectrum with the masses of all possible amino acid combinations. It takes advantage of the Hough transformation for spectra recalibration and outlier exclusion. The output is a list of the peptide fragments with their sequences and position in the protein.

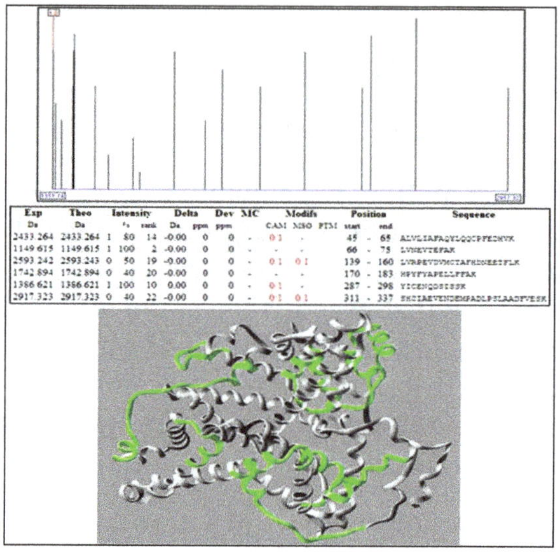

Figure above shows connection between the clinical laboratory and bioinformatics. Human serum albumin is decomposed into peptide fragments by trypsin. Mass spectrometric methods deliver for every protein a characteristic peptide mass fingerprint: the peptide mass spectrum (top). From this fingerprint, the peptide fragments can be identified by data base aided methods which enable an automatic comparison of the mass spectrum with the mass spectrum of all possible amino acid combinations (middle). A part of the identified peptide fragments (green) is visualized on the albumin 3D structure (bottom). The masses range from 1000 - 3000 Dalton.

Once the sequences of the peptide fragments have been determined they are linked together to the whole protein sequence. If the structure of a protein of known amino acid sequence has not yet been determined experimentally, it can be determined by homology modelling.

Homology Modelling

Because of the abundant number of known protein sequences, which exceed by far the number of experimentally determined structures, the relatively slow and expensive experimental methods have been supplemented by theoretical methods. Homology modelling is a form of protein tertiary structure prediction, based on the assumption that proteins that are homologous in sequence are similar in structure.

In homology modelling a protein sequence with an unknown structure is aligned with one or more protein sequences with known structures. The necessary condition for successful homology modelling is a detectable similarity between the amino acid sequences (more than 30%) allowing the construction of a correct alignment.

Principle

Homology modelling is a knowledge-based prediction of protein structures. The method uses parameters extracted from existing structures to predict a new structure from the sequence. This approach is possible because changes in the amino acid sequence usually cause only small structural changes. Therefore homology modelling first involves the finding of already known protein structures and then building the query sequence onto the homologous template structures. The steps in homology modelling are:

- The sequence with the unknown structure (the target) is used as a query to find similar sequences with known structures (the templates).
- The sequences are brought into an optimal alignment.
- The backbones of the templates are used to model the protein backbone of the target sequence.
- Loop-modelling procedures are used to fill gaps in the alignment.
- The side chains are added and their positions are arranged.
- The obtained structure is optimized by energy minimization or knowledge-based optimization.

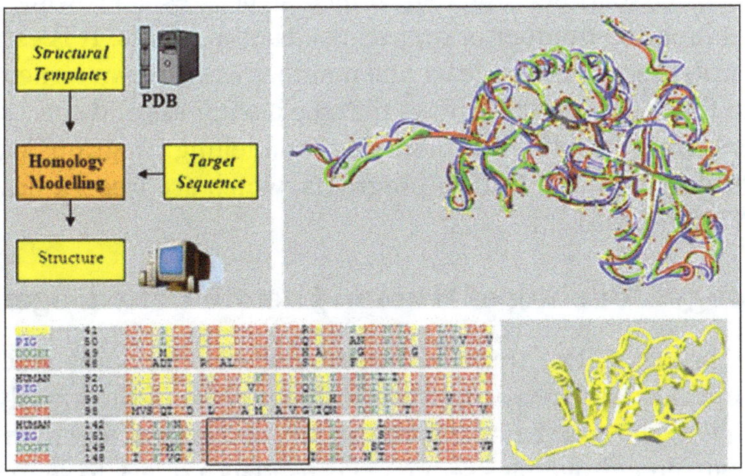

Figure above shows principles of homology modelling: In homology modelling a protein sequence with an unknown structure (the target) is used to find protein sequences with known structures (the templates). First the sequences are brought into an optimal alignment (bottom left). A part of the active site was determined by detecting the pig lactate dehydrogenase residues in the environment of the bounded pyruvic acid and highlighted (square) in the alignment. After the homology modelling procedure, the target sequence is "wrapped" around the superposed template structures (top right). The structure of the target is constructed by satisfying all the restraints, taken from the template structures, as well as possible (bottom right).

These steps are found in common in every method of homology modelling. A crucial issue in successful homology modelling is the quality of the alignment. The sequence alignment is used to determine the equivalent residues in the target and the template protein. Then the structure of the target protein is constructed by exploiting the information from the template structure. Some algorithms rely on rules of spacing between atoms, bonds lengths and angles etc. from observed values in known protein structures. These data are extracted from the template in form of spatial restraints and the structure of the target is constructed by satisfying all the restraints as well as possible. Other algorithms build the core of the model by averaging the backbone atom positions of the template structures, whereas the templates are weighted by the sequence similarity to the target sequence. The side chains are built by iso-sterically replacement of side chains in template structures. Of course, the predicted structure in homology modelling is not exactly the same as the template structures because of variations in the amino acid sequences. But structural deviations are allowed only in the range of differences found between homologous structures.

Distortions in the structure, which have been introduced by rigid modelling, are regulated by energy minimization of a molecular force field. Energy minimization is a technique that changes the conformation of a molecule in order to decrease its energy as much as possible. Energetically unreasonable features in conformations are for example: atoms that clashes with each other; unfavourable long bonds or bond angles etc. The predicted structure is optimized by energy minimization allowing the structure to settle into a lower energy conformation as similar as possible to the modelled conformation.

The chance to find a homologous sequence with a known structure in the PDB continually increases because the number of experimentally determined structures is growing rapidly and because physical constrictions limit the number of occurring structural folds. It is estimated that in a few years for every new protein sequence, at least one member with known structure of the same family will be available in the database. This shows that the usefulness and success of homology modelling is steadily increasing. The different homology modelling programs carry out single steps by proprietary algorithms. Two well-known programs, which are free for academic research, are MODELLER and SWISS-MODEL.

Example: Homology Modeling of Human Lactate Dehydrogenase

In order to illustrate the principles, homology modelling with the human lactate dehydrogenase sequence is carried out. Lactate dehydrogenase is the enzyme of the glycolysis which catalyses the reduction of pyruvic to lactate. Lactate is the salt of the lactic acid which can cause stiffness. First, the human lactate dehydrogenase sequence is used as a query to search the PDB database for homologous sequences of known structures. To this purpose, the program BLAST (Basic Local Alignment Search Tool) was used, where the query sequence is entered via a simple web form. This program performs pairwise comparisons of sequences. As a result, the sequences of lactate dehydrogenase from pig, mouse and dogfish were found as homologues. The alignment with the human lactate dehydrogenase sequence shows an E-value (expection-value) of e-142 for the sequence from pig-lactate dehydrogenase, e-136 from mouse-lactate dehydrogenase and e-132 from dogfish-lactate dehydrogenase. The E-value reflects the likelihood that a given sequence alignment is just random (occurred by chance). Alignments with low E-values are very significant, it means there is a high probability that the sequences are homologous. The three structures were selected as templates for the homology modelling. The output from the homology modelling process is also a structure file in PDB format, containing the atomic coordinates of the calculated structure.

Evaluation of the Predicted Protein Structures

The quality of the structure prediction can be evaluated with ProCHECK, a program that compares the geometry of the predicted structure with well-known high resolution structures. Important estimation criteria (stereochemical parameters) are: bond lengths; bond angles; torsion angles and planarity. The so called G-factor provides a measure of how realistic a given structure is. The G-factor is based on the observed distributions of the stereochemical parameters. A low G-factor indicates a low-probability conformation.

Applications of Molecular Modelling

Homology modelling remains, at present, the most efficient technique to provide accurate structural models of proteins. Typical uses of homology modelling are: designing of site-directed mutants to test hypothesis about function; identifying binding sites; improving inhibitors for a given binding site etc. The structures of medical relevant proteins involved in malaria, nicotinic acetylcholine receptor, bacteria inhabiting the gastrointestinal tract, anticancer drugs, hematopoietic stem/progenitor cell selection, autoimmunity and many more, have been determined by homology modelling.

Molecular Force Fields

Molecular force fields are used to calculate the energy of a given molecular structure. A task is the finding of the optimal geometry of stable molecules or of different possible confirmations. This can be done for example after a structure determination with homology modelling. The problem is then reduced to determine energy minima on the potential energy surface. Force fields are based on the observation that molecules are composed of units that are similar in different molecules. For example all C-H bonds have roughly constant lengths and their stretch vibrations are similar in different molecules. The empirical experience that molecules are composed of group of atoms, which are similar in a variety of molecules, is implemented in molecular force fields as concept of atom types. The atom type is defined by its chemical environment which can be distinguished by: the atomic charge; the nearest neighbours; the hybridization etc. The interactions are then calculated according to the atom type. Therefore force field methods are restricted to classes of molecules where information already exists. In molecular force fields the electronic energy is included as parametric function of the atomic coordinates where the parameters are fitted to experimental data. The building blocks in molecular force fields are atoms; electrons are not included as individual particles. The quantum aspects of the nuclear motion are neglected and the dynamics of the atoms in the molecule is treated by classical mechanics.

Formulation of Molecular Force Fields

The molecular force field describes a molecule as a collection of interacting atoms, with Cartesian coordinates r, that are expressed by simple analytical functions. The different interactions can be studied separately. Molecular force fields describe the potential energy surfaces. A general formulation of molecular force fields is as follows:

$$V=\sum_{bonds} K_r(r-r_0)^2+\sum_{bends} K_\Theta\left(\Theta-\Theta_0\right)^2+\sum_{dihedrals} V_n\left(1+\cos(n\Phi)\right)$$
$$+\sum_i\sum_j\left(\frac{A_{ij}}{r_{ij}^{12}}-\frac{B_{ij}}{r_{ij}^{6}}\right)+\sum_i\sum_j\frac{q_iq_j}{r_{ij}}+\sum_i\sum_j\left(\frac{C_{ij}}{r_{ij}^{12}}-\frac{D_{ij}}{r_{ij}^{10}}\right)$$

The first three terms describe the energetically situation of covalent bonds. The main parameters are: bond length ($r-r_0$), bond angle (Θ) and the dihedral angles (Φ). The first part describes the extension (stretching) of the covalent bonds, the second part the distortion of the bond angles, the third part the distortion of the dihedral angles of rotation of the bonded atom pair around the central bond). The constant expressions ($K_{r,}K_{\Theta,}V_n$) are dependent on the kind of covalent bond. The bond lengths are determined by the types of involved atoms and the number of the shared electrons between the atoms. An important covalent bond, contributing to protein structure stability is the disulphide bond between the sulphur atoms of two side chains of cystein residues. Bond angles are constrained by the structure of the electron orbital. The dihedral angles are constrained mainly by steric hindrance.

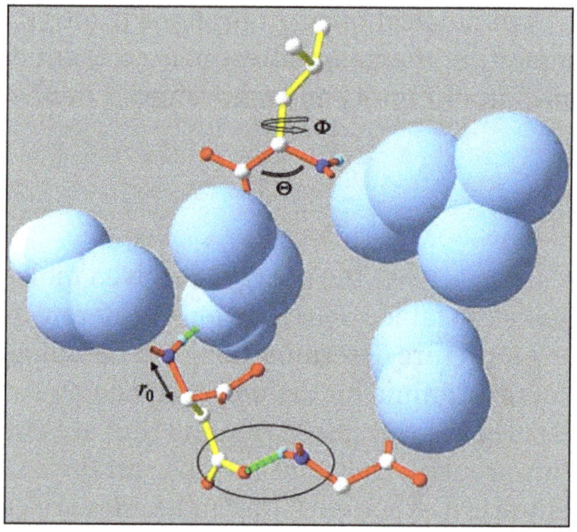

The figure above shows amolecular force field describes the energy of a molecular system with respect to: bond lengths; bond angles; van der Waals forces; electrostatic forces and hydrogen bonds. The energy of covalent bonds depends from bond length (r_0), bond angle (Θ) and the dihedral angles (Φ). Due to the stretching of bond lengths and the distortions of bond and dihedral angles, energetically unfavourable situations may occur in a protein molecule. Steric hindrance results from overlapping of the, with the residues associated, van der Waals spheres (blue), leading to strong repulsive forces. Hydrogen bonds (green dotted lines) are stabilizing forces arising between a proton donator and a proton acceptor. The peptide backbone is coloured in red, whereas the side chains are yellow.

The fourth term describes the van der Waals forces. Due to the movement of the electrons around the atomic nucleus, an atom can be considered as an electric dipole. The dipole of an atom polarises the neighbour atom resulting in a transient attractive force between the atoms. Conversely, at short ranges a repulsive force between the electrons of both atoms arises. The radius at which the repulsive force begins to increase sharply is called the van der Waals radius. Steric interactions arise when the van der Waals spheres of two non-bonded atoms are approaching and entering each other. The parameters A_{ij} and B_{ij} depend on the type of the involved atoms.

The fifth term describes the electrostatic forces arising between atoms carrying a positive ionic charge and atoms carrying a negative ionic charge. Charge-charge interactions between ions are called salt bridges. They occur between the side chains of the amino acids with opposite charges. They play a significant role in protein structure stabilization. There are other, weaker interactions that occur between dipolar residues (carrying partial positive and partial negative charge) and other partial or ionic charges: dipole-dipole, charge-dipole, charge-nonpolar and dipole-nonpolar.

The sixth term describes hydrogen bonds arising between a proton donator and a proton acceptor. These are weak chemical/electrostatic interactions between two atoms. They arise from the interaction of two polar groups containing a proton donor (amino group) and proton acceptor (carboxyl group). The donor group contains a hydrogen atom, covalently bonded with an electronegative atom, with partially positive charge. The electron of the proton is partially shifted to the bonding partner. The acceptor group has a partial negative charge with no attached proton. The parameters C_{ij} and D_{ij} depend on the types of the proton donator and acceptor atoms. The strength of hydrogen bonds depends on the distance and the bond angle. The overlapping of the electronic orbitals of the involved atoms in hydrogen bonds is usually small. Hydrogen bonds are identified if a proton donor atom and a proton acceptor atom of such groups are interacting at distance equal to or less than 3.6 angstroms. Hydrogen bonds can be formed between atoms of side-chains and/ or peptide backbone atoms. They are one of the most stabilizing forces in proteins (mainly the secondary structures like α -helices and β -sheets) and are responsible for the binding of the protein to substrates.

Applications of Molecular Force Fields

Energy minimization of the molecular force field establishes physically realistic structural configurations of molecules (for example after homology modelling). Equilibrium geometry of a molecule, with respect to bond lengths, angles, non overlapping van der Waals spheres etc., describes the atomic coordinates at a minimum on the potential energy surface. Different minima are connected with different conformations and saddle points represent transition states. Monte Carlo methods are used to simulate equilibrium properties and transitions between different states. For the study of molecular mechanics (MM), molecular force fields provide information about the time-dependent behaviour of molecules resulting in interconversions of different conformations and chemical reactions. Force field methods are however inherently not suitable to describe electron rearrangements and transfers in molecules, which must be done by quantum theoretical methods. An advantage of molecular force fields method is the speed with which calculations can be performed, enabling the application to large bio molecules. Even with moderate computer power, molecules with thousands of atoms can be optimized. This facilitates the molecular modelling of proteins and nucleotide acids which is nowadays done by most pharmaceutical companies.

There are several common force fields in use in professsional molecular modelling programs. They are simplified for the treatment of macromolecules, for example, by not considering hydrogen atoms explicitly (united atom approach). The force fields are calibrated against experimental data like: bond lengths, bond angles etc. Common molecular force fields are: GROMOS, CHARMM, AMBER, MM+, BIO+ and OPLS force fields.

Structural Alignments and Homologies

Similar protein structures are determined by direct comparison of 3D protein structures with the VAST (Vector Alignment Search Tool) algorithm. The neighbour structures are often examples of remote homology, undetectable by sequence comparison. As such they may provide insights into structure, function and evolution of a protein family. The principle of VAST's significance calculation is based on the definition of the unit of tertiary structure similarity as pairs of secondary structure elements (SSE) that have similar type, relative orientation, and connectivity. Similarities of small substructures that occur by chance are not considered. VAST detects via structural alignment similarities between proteins that cannot detected by sequence alignments because of poor similarities between the sequences.

To compare two proteins, the structures are superposed. The difference between two protein structures is expressed by the root mean squared deviation (RMS) of the respective atomic positions in the two structures.

$$RMS = \sqrt{\frac{\sum_i d_i}{N}}$$

d_i is the distance between two corresponding atoms and N is the number of considered atoms. Mostly the differences of the positions of the respective C_α atoms of the backbones in the two structures are measured. RMS values up to 1 angstrom show a high similarity between the two structures. The protein structures are described in Cartesian coordinates and each structure has a build-in orientation in its proper coordinate system. To compare the two structures, one structure serves as reference and the other structure must be superimposed. This allows the evaluation of which parts of the structures are showing a good or high similarity and where significant structural deviations are located allowing to examine the patterns of structural conservation and change within a protein family.

Protein Interactions

Processes inside and outside cells are described as networks of interacting proteins. The amino acids on the surface of a protein interact with other molecules such as substrates, ligands, receptors etc. The shape of a protein is of special importance for intermolecular interactions. These interactions can be described in terms of locks and keys. To enable an interaction, the shape of the lock (for example the enzyme) must be complementary to the shape of the key (the substrate). In the immune system the shape of the antibodies must be complementary to the shape of the antigens to initiate an immune response. The body develops antibodies with the right shape to attack specific antigens (for example a virus). Therefore the shape of the molecular surface of a protein offers information on how two proteins in a metabolic pathway interact with each other or why an enzyme is specific to a particular substrate.

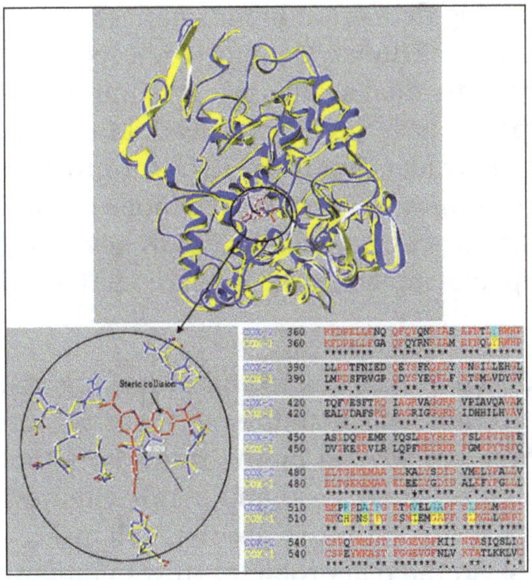

Figure above shows the structure of COX-2 (blue) is superimposed with the structure of COX-1 (yellow). RMS=0.93. The structural alignment (only a part is displayed) shows a high portion of similar residues in both chains and the corresponding residues of the active sites. The active centre, where COX-2 bounds the selective inflammation inhibitor SC-558 (red), is shown. The interchanging of valine by isoleucine at position 523 leads to a steric collision with SC-558 in the case of COX-1.

Experimental Determination of Protein-Protein Interactions

The yeast two-hybrid (Y2H) method was developed to study experimentally the protein-protein interactions. The principle of the method is as follow: In a medium without, for example, histidine, yeast can only grow if the corresponding gene (reporter gene) for the production of histidine is switched on. First, the transcription factor responsible for the expression of the reporter gene is cut into 2 parts: the DNA binding domain and the activation domain which stimulates the RNA polymerase to begin transcription. The protein of interest (protein A) is fused with the DNA binding domain part. Its potential counterpart (protein B) for the protein-protein interaction is fused with the activation domain. If both proteins are interacting they form together with the DNA binding and activation domain an active promoter complex which initiates the transcription of the reporter gene and yeast is growing (if they don't interact, the reporter gene stays inactive). This situation occurs when 2 yeast stems, where each carries one of the two fusion proteins on a plasmid, are merged (mating).

Integrative Bioinformatics

Integrative bioinformatics is an emerging field in the big data era, offering a steadily increasing number of algorithms and analysis tools. However, for researchers in experimental life sciences it is often difficult to follow and properly apply the bioinformatical methods in order to unravel the complexity and systemic effects of omics data.

We are now increasingly in the big data era. Bioinformatics is facing much more heterogeneous biological data with huge volumes. “Human Genome Projects” like “one million HGP” are going on, leading to more and more individual sequences. It is not only for human, but also for other species, as more and more species have been sequenced. Moreover, it does not simply measure whole tissue samples, but distinctly identify DNAs/RNAs or proteins at a cellular level. Single cell sequencing and single cell proteomics will generate millions of profiling datasets in a short time period. The multi-omics data brings us new challenges to develop appropriate integrative bioinformatics approaches to manipulate integrate and model complex biological systems at spatial and temporal scales.

Since biological data are subjective and biased, often lacking standardization and reproducibility, and some databases are not well maintained, these resources are becoming more and more degraded. Although there are several bioinformatics methods developed to deal with a certain problem, often only one is widely used and highly cited, which encourages becoming a common/standard method. In many cases, we are not well aware of the original hypothesis of such methods, which may mislead the real problem. How to integrate the multi-omics data with different biological/technical conditions and bias? How to share/deposit data under an acceptable intelligence and ethic policy? Is our traditional data mining and machine learning methods suitable for big data? More powerful tools for multiple scale biological interactome modeling and simulation. How to uncover hidden patterns from such a huge and heterogeneous amount of omics data and allowing the creation of predictive models for real-life applications? Nevertheless, advances in biological technologies and computational methodologies have provided a huge impetus. There are several directions which may lead to break the bottleneck of Integrative Bioinformatics:

- Integration of multiple biological data towards systems biology. Different omics data is reflecting different aspects of the biological problem. For instance, previously biological networks are regarded as gene regulatory network, protein-protein interaction network and metabolic networks. Now we know that non-coding RNAs, including microRNAs, siRNAs, lncRNAs, ceRNAs and cirRNAs etc. play more important roles in regulations. Therefore, an integrative interactome model (e.g. a virtual cell) of known parts and non-coding RNAs need to be built.

- Integration of various bioinformatics methods and approaches. Often, to solve a problem, there are many different methods developed by many groups. These methods may perform differently, some good, some bad. However, individual results are often unreliable. In particular cases, the often-used methods may be unreliable or simply ineffective. It is suggested to depend on a variety of results by all methods. With various methods, we are able to integratively develop tailored bioinformatics pipelines to facilitate better understanding of biological problems.

- To integrate multiple biological data and different methods/approaches, well developed traditional data mining methods such as NN, SVM, HMM are available. However, they are not good enough to deal with high dimensional omics data and big data sets. So far, deep learning methods such as CNN, RNN have been used. Combing with big data, and other approaches, artificial intelligence (AI) has been successfully applied in bioinformatics, especial in the field of biomedical image analysis.

- Computing infrastructure development. Integrative Bioinformatics in the big data era requires a more advanced IT environment. To facility the related computing and visualization demands, both hardware (e.g. GPU) and software (e.g. Tensorflow) are developing. Supercomputers are used. Cloud services are provided by more and more institutes and big companies.

Industry Aspects

During the turn of the century the availability of the fully sequenced human genome and other model organisms sparked a boom of bioinformatics companies aiming to address the challenges in medicine, plant and other life sciences using computational methods. Despite initial successes like improving genome annotations or modeling of more complex protein structures, big promises like in-silico drug discovery were not able to be kept and even huge players like Lion Biosciences diminished. Nevertheless, the enthusiasm and learning of that time led to the establishment of dedicated bioinformatics functions within almost all of life sciences industries. These bioinformatics functions would be placed within the R&D functions of life science companies. As dedicated talent in bioinformatics were rare, biologists, computational scientists or even physicists and others strained in the new area of bioinformatics. The demand of industry for talent influenced the academic world and drove the creation of more and more bioinformatics or related curricular.

As such bioinformatics functions are embedded in the whole R&D ecosystem of a life science company there are already surrounding data systems in-place concerning relevant bioinformatics data domains. These data systems can range from simple spreadsheets used by the scientists, to Access databases and relational database systems. Understanding the data stored in these systems and adding the contributions of bioinformatics tools and predictions to the R&D ecosystem heavily relies on integrative bioinformatics approaches. Breaking up data silos between functional units within the R&D ecosystem is a prerequisite to drive not only track and traceability of processes but also the discovery of new insights. Technologies like semantic web or linked data provide the base infrastructure of efficient bioinformatics functions. Ontologies either reused from public repositories or customized together with R&D scientists establish a common language, which should also be machine interpretable. FAIR (findable, accessible, interoperable, reusable) principles of data management are more and more being adopted in industry.

Approaches

Semantic Web Approaches

In the Semantic Web approach, data from multiple websites or databases is searched via metadata. Metadata is machine-readable code, which defines the contents of the page for the program so that the comparisons between the data and the search terms are more accurate. This serves to decrease the number of results that are irrelevant or unhelpful. Some meta-data exists as definitions called ontologies, which can be tagged by either users or programs; these serve to facilitate searches by using key terms or phrases to find and return the data. Advantages of this approach include the general increased quality of the data returned in searches and with proper tagging, ontologies finding entries that may not explicitly state the search term but are still relevant. One disadvantage of this approach is that the results that are returned come in the format of the database of their origin and as such, direct comparisons may be difficult. Another problem is that the terms used in tagging and searching can sometimes be ambiguous and may cause confusion among the results. In addition, the semantic web approach is still considered an emerging technology and is not in wide-scale use at this time.

One of the current applications of ontology-based search in the biomedical sciences is GoPubMed, which searches the PubMed database of scientific literature. Another use of ontologies is within databases such as SwissProt, Ensembl and TrEMBL, which use this technology to search through the stores of human proteome-related data for tags related to the search term.

Some of the research in this field has focused on creating new and specific ontologies. Other researchers have worked on verifying the results of existing ontologies. In a specific example, the goal of Verschelde, et al. was the integration of several different ontology libraries into a larger one that contained more definitions of different subspecialties (medical, molecular biological, etc.) and was able to distinguish between ambiguous tags; the result was a data-warehouse like effect, with easy access to multiple databases through the use of ontologies. In a separate project, Bertens, et al. constructed a lattice work of three ontologies (for anatomy and development of model organisms) on novel framework ontology of generic organs. For example, results from a search of 'heart' in this ontology would return the heart plans for each of the vertebrate species whose ontologies were included. The stated goal of the project is to facilitate comparative and evolutionary studies.

Data Warehousing Approaches

In the data warehousing strategy, the data from different sources are extracted and integrated in a single database. For example, various 'omics' datasets may be integrated to provide biological insights into biological systems. Examples include data from genomics, transcriptomics, proteomics, interactomics, and metabolomics. Ideally, changes in these sources are regularly synchronized to the integrated database. The data is presented to the users in a common format. Many programs aimed to aid in the creation of such warehouses are designed to be extremely versatile to allow for them to be implemented in diverse research projects. One advantage of this approach is that data is available for analysis at a single site, using a uniform schema. Some disadvantages are that the datasets are often huge and difficult to keep up to date. Another problem with this method is that it is costly to compile such a warehouse.

Standardized formats for different types of data (ex: protein data) are now emerging due to the influence of groups like the Proteomics Standards Initiative (PSI). Some data warehousing projects even require the submission of data in one of these new formats.

Other Approaches

Data mining uses statistical methods to search for patterns in existing data. This method generally returns many patterns, of which some are spurious and some are significant, but all of the patterns the program finds must be evaluated individually. Currently, some research is focused on incorporating existing data mining techniques with novel pattern analysis methods that reduce the need to spend time going over each pattern found by the initial program, but instead, return a few results with a high likelihood of relevance. One drawback of this approach is that it does not integrate multiple databases, which means that comparisons across databases are not possible. The major advantage to this approach is that it allows for the generation of new hypotheses to test.

Translational Bioinformatics

Translational Bioinformatics is the development of storage, analytic, and interpretive methods to optimize the transformation of increasingly voluminous biomedical data, and genomic data, into proactive, predictive, preventive, and participatory health. Translational bioinformatics includes research on the development of novel techniques for the integration of biological and clinical data and the evolution of clinical informatics methodology to encompass biological observations. The end product of translational bioinformatics is newly found knowledge from these integrative efforts that can be disseminated to a variety of stakeholders, including biomedical scientists, clinicians, and patients.

Currently, the transformation of huge volumes of complex data into clinically useful knowledge requires the convergence molecular bioinformatics, biostatistics, statistical genetics and clinical informatics into computational methods to establish a better/more practical, and accelerated, path from discovery to improving patient care, and important discoveries for practical use. In other words, achieving these goals involves interdisciplinary collaboration, and one that is essential to advancing translational genomics, similar to those used in large "big data" research projects.

The importance of translational bioinformatics may be best understood in the things it is teaching us, things not previously knowable. For example, it is identifying flawed science, improving estimates of relative pathogenicity of human genetic variants, inferring new insights about underlying genetic mechanisms of disease, and identifying promising new drug indications based on curating large volumes of scientific literature. While, sequencing an exome for a clinical diagnosis can be a routine task, the interpretation of the data to make an actual diagnosis or treatment plan is much more complex. Out of the many thousands of variants identified, many of them will have to be evaluated for their clinical utility. At times, for perhaps a simple Mendelian disorder this may be as simple, as only a single variant will need to be identified and considered. But for more complex diseases (e.g. cancers, diabetes, or neurodegenerative diseases) multiple variants will need to be identified. It is only by asking the correct questions about the patient and the disease, along with employing the right computational tools that correct answers can be achieved. In some cases appropriate questions and tools already exist and can be applied, while in other cases, they still need to be developed. Thus, the translational bioinformatics field is poised to address these issues and help generate the right answers.

Determining the right question to ask is now a collaborative effort involving non-conventional sources. In the process, unaddressed areas or unchallenged assumptions can be identified and so researched. Bioniformaticians can, and for that matter do, create new tools to answer the questions, pioneer and validate their use, and report findings demonstrating the tool's ability to provide good answers. Thus, the field has moved from mere doing to also questioning and therefore to being integrally involved in research design. In this way innovative bioinformatics methods are revolutionizing translational science. Crowd based discovery, new methods for using genome/transcriptomes to recommend cancer treatments and more trained systems, such as machine learning, are but some examples. Furthermore, it is our increasing ability to directly investigate those biological responses in real time is enabling the field to understand the complexity of disease processes.

Thus, with the right questions and the right tools to answer them, translational bioinformatics can translate DNA sequences into new discoveries, novel diagnostics and fundamental causes of disease as targets for new therapeutics. The scope of application, though, goes beyond the biomedical sciences to health care delivery, health economy and health policy. In this way, translational bioinformatics is crucial to advancing translational genomics (omics).

Big Data and Biomedicine

As technology enables us to take an increasingly comprehensive look across the genome, transcriptome, proteome, etc., the resulting datasets are increasingly high-dimensional. This in turn requires a larger number of samples in order to achieve the statistical power needed to detect the true signal. The past decade or so has seen an increasing number of large-scale bio-repositories intended for clinical and translational research all over the world. These projects comprise both information and biospecimens from individual patients, enabling researchers to reclassify diseases based on underlying molecular pathways, instead of the macroscopic symptoms that have been relied on for centuries in defining disease. These various projects involve different models of participation, ranging from explicit informed consent to use of de-identified biospecimens and their associated clinical information from EHRs (also de-identified). The informed consent is the most ethically rigorous model, but also the most expensive. The use of de-identified specimens and data is more scalable, and financially feasible. However, as complete genomic data are increasingly used, it is impossible to truly de-identify these data. This raises ethical issues regarding patient privacy and data sharing. In the United States, legislation known as the "Common Rule" addresses these issues. In 2015, a notice of proposed rule-making was released to solicit feedback on some major revisions to the law, which was originally passed in 1991. Much has changed within biomedical research in the intervening years.

In order to accrue the numbers of samples required for the "big data" discipline that biomedical research is becoming, the ability to use patient data and samples in research would be of significant benefit. One major point addressed in the aforementioned NPRM is the ability for patients to give broad consent for future use of data and samples, without knowing the specifics of research studies ahead of time. As we move toward the learning healthcare system (LHS) model in which every encounter is an additional data point, explicit research registries will become less relevant. They will be too expensive to maintain, and larger numbers of patients/participants will be available through federated initiatives that allow a researcher to query across institutions regionally, nationally, and even internationally. The National Patient-Centered Clinical Research Network (PCORnet) takes this approach, enabling clinical outcome research through federated pragmatic clinical trials. Importantly, this initiative emphasizes partnership with patients and their advocates, so that they are empowered as collaborators, with a say in what research questions matter most.

LHS is about moving from evidence-based practice, i.e., clinical care decisions based on conscientious use of current best evidence, to practice-based evidence, i.e., the generation of evidence through collection of data in the real-world as opposed to the artificially-controlled environment of randomized clinical trials. In recent decades, the biomedical enterprise has strived to practice medicine in a way that is supported by the best possible evidence from randomized clinical trials. But clinical trials have their own issues. They are expensive, and they tend to be very different from real life scenarios. Criteria for inclusion in a trial often include the absence of common comorbidities or use of common medications. Compliance tends to be high, but the cohort being studied is often not representative of the target population for the treatment in question. In LHS, translation becomes bi-directional. Research is used to inform practice, whereas data that are generated in the course of clinical care can in turn be used for both hypothesis generation and validation through pragmatic trials. Data derived from clinical care can thus inform clinical guidelines and future practice.

Secondary Use

Secondary use of data refers to data that are created or collected through clinical care. In addition to use in caring for the patient, these data may also be crucial for operations, quality improvement, and comparative effectiveness research. Some assert that the term "secondary use" should give way to the term "continuous use." They argue against the notion that data collected at the point of care are solely for clinical use, and everything else is secondary. We should be maximally leveraging this valuable information. Nonetheless, there is a legitimate concern about data quality. Data in the EHR are often sparse, incomplete, and even inaccurate. This makes these data wholly unsuitable for certain purposes, but still sufficient for others.

TBI tends to focus on molecules, newly accessible in high dimensions based on novel high-throughput technologies. Phenotyping is a closely-related challenge, more complex than it might seem. Disease is not binary: even within a very specific type of cancer, a tumor's genomic profile may be quite different among the precise sampling locations and sizes . There are a number of groups focusing on this problem: the Electronic Medical Records and Genomics (eMERGE) Network, the NIH Collaboratory, PCORnet, and the MURDOCK Study, among others. The Phenotype KnowledgeBase website is a knowledge base of phenotypes, offering a collaborative environment to build and validate phenotype definitions. The phenotypes are not (yet) computable, but it serves as a resource for defining patient cohorts in specific disease areas.

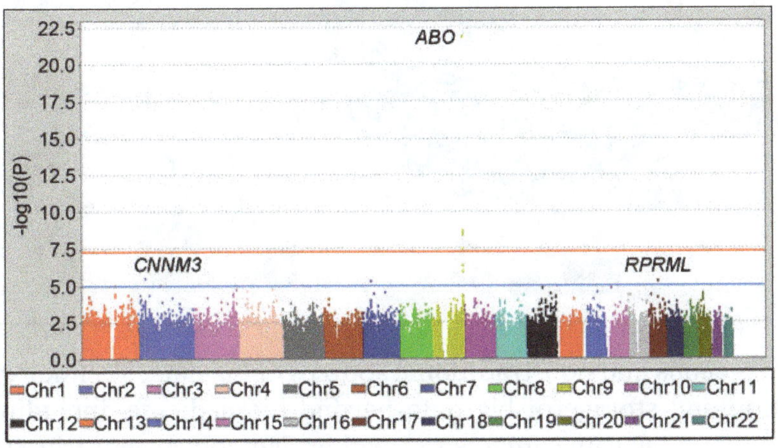

Figure shows a PheWAS Manhattan plot for a given SNP. This plot shows the significance of association between SNP rs965513 and 866 different phenotypes. Along the X axis different disease groups are shown in different colors. This is in contrast to an analogous plot for GWAS in which the X axis would represent the different chromosomes. The Y axis reflects the P value for each phenotype. Blue and red horizontal lines represent P value of 0.05 and Bonferroni corrected P value of $5.8 \times 10^{-5,}$ respectively. PheWAS, phenome-wide association studies; GWAS, genome-wide association studies. The figure was reproduced from with permission from Elsevier.

Instead of looking at the entire genome, PheWAS evaluates the association between a set of ge-netic variants and a wide and diverse range of phenotypes, diagnoses, traits, and/or outcomes. This analytic approach asks, for a given variant, do we see an enrichment of a specific genotype in any of these phenotypes? Figure below illustrates results using this approach. In standard GWAS analyses, the different color bands at the bottom represent the different chromosomes. In the case of PheWAS, they are different disease areas, e.g., neurologic, cardiovascular, digestive, and skin. Pendergrass et al. used a PheWAS approach for the detection of pleiotropic effects, where one gene affects multiple different phenotypes. They were able to replicate 52 known associations and 26 closely-related ones. They also found 33 potentially-novel genotype–phenotype associations with pleiotropic effects, for example the GALNT2 SNP that had previously been associated with HDL levels among European Americans. Here they detected an association between GALNT2 and hy-pertension phenotypes in African Americans, as well as serum calcium levels and coronary heart disease phenotypes in European Americans.

Another aspect of big data in biomedicine is the use of non-traditional data sources. A complex and detailed figure below showed various data types that could be mined for their effects on human health: weather, air traffic, security, cell phones, and social media among others. But strikingly the list did not include personal activity trackers, e.g., Fit Bit, Jawbone, or even the Apple watch. This omission of such a popular technology today is indicative of what a fast-moving field this is.

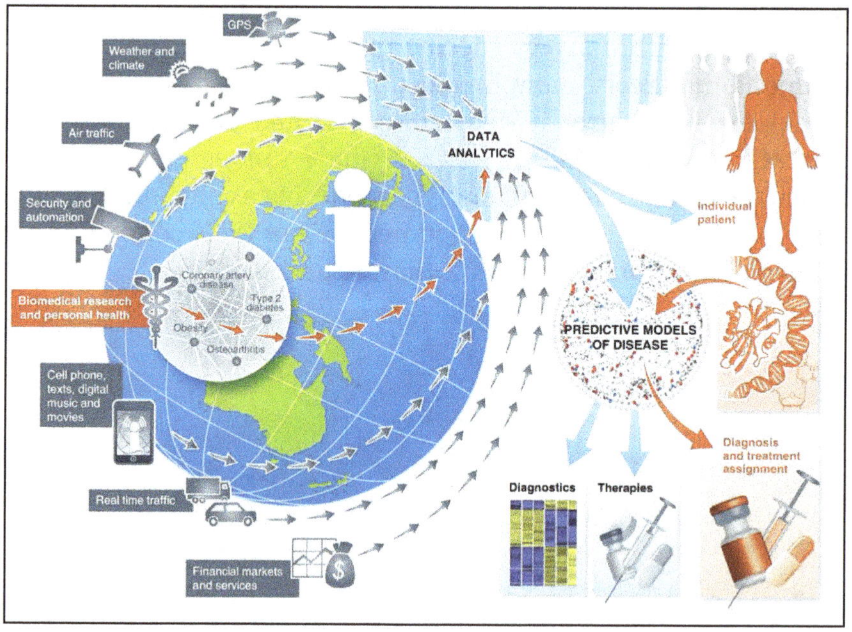

Figure shows heterogeneous and non-traditional sources of big data. Technological advances have enabled the collection and storage of big data beyond biomedicine, including everything from credit card transactions to security cameras to weather. Notably absent from this 2012 figure are smart watches and fitness tracking devices, which became pervasive in the years that followed.

Genomics in Clinical Care

One sees a number of examples of how genomic data are used in clinical care in the context of pharmacogenomics. But molecular data, and genomic data derived from next-generation sequenc-ing (NGS) in particular, have been used in a number of other contexts as well. One example took place at Stanford's Lucile Packard Children's Hospital, where a newborn presented with a condition known as long QT syndrome. In this specific case, the manifestation was unusually severe—the baby's heart stopped multiple times in the hours after its birth. Long QT syndrome can be caused by mutations in a number of different genes. It is necessary to know which gene harbors the mutation in order to know how to treat the condition. In this case, a whole-genome sequencing (WGS) was performed enabling identification of a previously-studied mutation, as well as a novel copy number variation in the TTN gene that would not otherwise have been detectable through targeted genotyping alone. Moreover, NGS enabled the answer to be obtained in a matter of hours to days instead of weeks.

Another example of DNA sequencing in clinical care involved diagnosis of infectious disease. A 14-year-old boy with severe combined immunodeficiency (SCID) had been admitted to the hospital repeatedly. He had headache, fever, weakness, nausea, and vomiting. His condition continued to decline to the point where he was put into a medically-induced coma. A normal diagnostic work up was "unrevealing" and the doctors were unable to determine the etiology of his condition. The patient was enrolled in a research study for pathogen detection and discovery in hospitalized patients. The protocol for this study involved performing NGS on the subject's spinal fluid. The results included detection of 475 reads corresponding to leptospira infection. Of note, the nor-mal test for leptospirosis involves detection of the patient's antibody response to the infection. In this case, the patient's SCID status prevents such a response, so the infection was not detectable through standard means.

Cancer has been one of the most active areas for the translation of genomic discoveries into changes in clinical care. One of the biggest players in this space is Foundation Medicine, which makes the Foundation One test, a targeted panel that uses NGS to test all genes known to harbor mutations in solid tumors. Johnson et al. looked retrospectively at approximately 100 patients who had undergone FoundationOne, and found that 83% of them had potentially actionable results from that test, of which 21% received genotype-directed treatment. Explanations for why the indicated treatment was not given in 79% of those cases included the decision to use standard therapy and clinical deterioration. These results indicate that application of genomic technology has transcended the research domain. In many cases, the findings are clinically actionable. Mirroring this fact, it is worth noting that some medical insurance companies in the United States started to offer coverage of the FoundationOne test in late 2014.

In addition to targeted panels, deep cancer sequencing can help shed light on drug sensitivity and resistance. Wagle et al describe a case of a 57-year-old woman with anaplastic thyroid cancer. Her specific tumor was initially sensitive to everolimus, a mammalian target of rapamycin (mTOR) inhibitor. Her doctors were able to sequence the tumor before it became resistant, revealing a mutation in TSE2, which encodes a negative regulator of mTOR. Normally mTOR is down-regulated by TSE2, but the mutation caused TSE2 not to down-regulate mTOR to a sufficient level. Therefore, everolimus, which inhibits mTOR, was effective in treating this specific cancer. Later this patient's cancer became drug-resistant, whereupon the newly drug-resistant part of the tumor was sequenced again. It was discovered that an mTOR mutation had cause mTOR not to be inhibited by allosteric inhibitors like everolimus. An allosteric inhibitor binds to the protein in question somewhere other than on the active site of the molecule. Knowing the specific cause for the newly-acquired resistance leads to other treatment options. They were able to switch to mTOR kinase inhibitors, in order to down-regulate the pathway through other mechanisms.

Despite these compelling cases, it is worth noting that at this point, tumor sequencing is the exception in clinical oncology, and not one of the routine procedures. Only after "first line" treatment has failed are tumors sequenced, and even that is largely confined to large academic medical centers. It will be interesting to see if, when, and how that changes.

While blood clotting and cancer are the areas where most actionable pharmacogenomics findings have been made, a notable exception is work by Tang and his colleagues, in which they describe the identification of a genotype-based treatment for T2DM with an alpha2A-adrenergic receptor antagonist. A specific genetic variant causes over-expression of the alpha2A-adrenergic receptor, and impaired insulin secretion. They hypothesized that if they could block the over-expressed receptor, they could increase insulin secretion. Those with the mutation showed a dose response to the drug as measured through levels of insulin. Participants without the mutation showed no such response. This research is especially interesting because it goes beyond cancer and blood thinners, into a chronic disease that is affecting an increasing portion of the world's population.

One of the earliest and best known DNA sequencing success stories was that of Nic Volker, who at age 2 developed severe gastro-intestinal issues resembling Crohn's disease, for which a diagnosis could not be determined. His story was covered in the Milwaukee Journal Sentinel in 2011 in a piece that was later awarded a Pulitzer Prize for explanatory reporting. After multiple life-threatening situations and a protracted diagnostic odyssey, Volker's genome was sequenced to look for a causal mutation. Doing so enabled the discovery of a mutation in the gene X-linked inhibitor of apoptosis (XIAP). Equally important, there existed a known therapeutic intervention for disorders caused by XIAP mutations. A progenitor cell transplant was performed and the patient's condition improved. Though Nic is not without ongoing health challenges, he celebrated his 11th birthday in October, 2015.

There is also reason for caution regarding use of NGS sequencing in clinical care. Dewey et al. performed Illumina-based WGS on 12 different participants. Confirmatory sequencing was formed on 9 of those participants by Complete Genomics Inc. Their findings included the fact that 10%–19% of inherited disease genes were not covered to accepted standards for SNP discovery. For the genotypes that were called, concordance between the two technologies for previously-described single nucleotide variants was 99%–100%. However, for insertion and deletion variants, the concordance rate was only 53%–59%. Approximately 15% were discordant, and approximately 30% of the variants could not be called by one technology or the other. In addition, inter-rater agreement for whether findings should be followed up clinically was only 0.24 ("fair") by Fleiss' kappa metric. Rater agreement was even worse for cardio-metabolic diseases, with the rate at which the two experts agreed on the need for clinical follow-up worse than random in those cases. Notably, the estimated median cost for sequencing and variant interpretation was about US $15,000, plus the price of the computing infrastructure and data storage. This means the cost of interpretation is significantly more than the proverbial US $1000 genome goal, but is also significantly less than the US $100,000 or $1 million some had feared.

Omics for Drug Discovery and Re-purposing

Much has been said about the protracted process involved in getting a drug through the FDA approval pipeline. Estimates are that the process can take on average 12 years between lead identification and FDA approval. This makes the prospect of drug repurposing an appealing one. Drug repurposing refers to taking an existing, FDA-approved compound and using it to treat a disease or condition other than the one for which it was originally intended. In the past, inspiration for this type of "off label use" has been largely serendipitous. For example, Viagra was initially aimed at treating heart disease, and turned out to be useful for erectile dysfunction. By using a pre-approved compound, early phase clinical trials can be avoided, which can save significant time and money.

One is to look for molecular signatures in disease and compare those to signatures observed in cells, animal models, or people who have been treated with different drugs. If anti-correlated signatures can be identified between diseases and drugs, administration of that drug for that disease may help cure the condition, or at least to alleviate the symptoms. One of the prominent early examples of a computational approach to drug repositioning was the Broad Institute's Connectivity Map (CMap). They made the data available as a resource intended to enable the identification of functional connections between drugs, genes, and diseases. Another example is work by Jahchan et al., in which they identified anti-depressants as potential inhibitors of lung cancer. Numerous disease and drug profiles and found an anti-correlation between gene expression seen after administration of anti-depressants and the pattern of expression observed in small cell lung cancer. They next transplanted these types of tumors into mice and found that with the drug, those tumors either shrunk or didn't even grow. They also used indigenous tumors in the mouse model for this type of cancer and found that the anti-depressant had promising results on those tumors and for other endocrine tumors as well. They were able to start a clinical trial, much faster than would have been possible by chance or by various other traditional methods, though unfortunately that trial was ultimately terminated for lack of efficacy. A similar approach was used by Sirota et al. to identify the anti-ulcer drug cimetidine as a candidate agent to treat lung adenocarcinoma and validate this off-label usage in vivo using an animal model of the lung cancer.

For example, Nygren et al. screened 1600 known compounds against 2 different colon cancer cell lines. They used Connectivity Map data to further evaluate their findings, and identified mebendazole (MBZ) as having potential therapeutic effect in colon cancer. Finally, Zhu et al. mined data from PharmGKB and leveraged the web ontology language (OWL) to perform semantic inference. They were able to identify potential novel uses and adverse effects of approved breast cancer drugs.

Personalized Genomic Testing

The year 2008 saw the founding of several companies that offered direct-to-consumer (DTC) genetic testing, reporting on a variety of genes for both health and recreational purposes. As of 2016, 23andMe was the last major player standing in the United States, with other companies having been acquired and/or changing their business models.

DTC genetic testing raises a number of interesting ethical, legal, and social issues. For several years, there was an open question as to whether or not these tests should be subject to government regulation. In November 2013, the US FDA ordered 23andMe to stop advertising and offering their health-related information services. The FDA considered these tests to be "medical devices" and as such to require formal testing and FDA approval for each test. In February 2015, it was announced that the FDA had approved 23andMe's application for a test for Bloom syndrome and in October 2015 it was announced that the company would once again be offering health information in the form of carrier status for 36 genes. Note that a 23andMe customer is able to download his or her raw genomic data and to use information from other websites to interpret the results, including Promethease, Geneticgenie, openSNP), and Interpretome for health-related associations.

Another important question raised by DTC genetic testing include whether the consumers are ready for this information. Traditionally, patients receive troubling health-related information in a face-to-face conversation with their doctor. There is some concern that patients are not competent or well-equipped to receive potentially distressing news through an Internet link. To help mitigate this concern, 23andMe "locks" certain results, making them accessible only if the user clicks through an additional link, indicating they truly want to know.

What about the healthcare providers? Are they ready to incorporate genomic data, patient-supplied or otherwise, into treatment decisions? In a case described in 2012, a 35-year-old woman informed her fertility care provider that her 23 and Me results revealed a relatively common (1 in 100) blood clotting mutation. She was surprised when her provider responded by saying that, were she to become pregnant, she would need to be put on an injectable anti-coagulant throughout her pregnancy. With no family history of blood clotting disorders, nor personal history of recurrent miscarriage, this mutation would have gone untested, had it not been for the DTC results. However, when this patient did become pregnant and consulted with a specialist whose expertise was in blood clotting disorders in pregnant women, the anti-coagulant was indeed prescribed. Note that the guidelines have since changed, and the prophylactic treatment for clotting would not be prescribed today. It is also worth noting that in the United States, the Genetic Information Nondiscrimination Act (GINA) prevents employers and health insurers from discriminating against anyone based on their genetic information. It does not, however, cover long-term care, disability, or life insurance. Therefore when this woman applied for life insurance after her twins were born, the rate she was offered was more than twice what it would have been had she not known about the blood clotting mutation and been treated prophylactically for the increased risk it conferred.

A more positive example of where genetic testing is helping patients is a case presented at the American Neurological Association conference in 2014. A patient had a history of Alzheimer's disease on her mother's side of the family. She did not know if she was a carrier, nor did she want to know. But she wanted to ensure that she did not pass that mutation to her future children. Preimplantation genetic diagnosis (PGD) testing enabled her doctors to select embryos that did not have that Alzheimer's disease gene mutation. The patient herself was never tested, nor was she informed how many (if any) of the embryos contained the mutation.

Privacy

The researchers used the short tandem repeat (STR) data from the Y chromosome, year of birth, and state of residence, combined with information from public genealogy websites to identify individuals. They did this by starting with publicly-available STRs and entering them on genealogy websites to identify matches. Note that their accuracy was not 100%. Though they were able to identify Craig Venter based on his genomic data, they failed to identify several other individuals, particularly those with more common last names. Overall, they reported a 12% success rate in recovering surnames of US males. They were also able to reconstruct Utah family pedigrees based on 1000 Genomes Project data and other publicly-available sources. Due to a number of cultural and historical factors, families in Utah tend to be large and genealogically well-documented, and an unusually high proportion of the Utah population has participated in scientific studies involving genomic data. It is worth noting that the researchers did not release any names that were not previously public, nor did they use the information for any nefarious purposes. Their primary interest was to demonstrate that such re-identification was possible. Interestingly, and somewhat surprisingly, the terms of use for the datasets did not prohibit re-identification.

References

- Overview-of-Bioinformatics: inflibnet.ac.in, Retrieved 09, March 2020
- Structural-Bioinformatics-From-the-Sequence-to-Structure-and-Function-228909583: researchgate.net, Retrieved 12, June 2020

Chapter 2

Sequence Analysis

Sequence analysis refers to the way in which proteins or DNA sequences are subjected to a wide range of analytical methods to understand its features, functions, structure or evolution. This chapter deals with sequence analysis and its various aspects like DNA sequencing, sequence assembly, sequence alignment, multiple sequence alignment, etc. in an in-depth manner for the benefit of the reader.

Sequence analysis is a term that comprehensively represents computational analysis of a DNA, RNA or peptide sequence, to extract knowledge about its properties, biological function, structure and evolution. To carry out sequence analysis efficiently, it is important to first understand the source of the data, i.e., the different experimental methods used for determining the biological sequence. We then need to follow analytical strategies, depending on whether the sequence is genomic, transcriptomic or proteomic. Databases currently warehousing the enormous data on these biomolecules will need to be first checked for the presence of similar sequences, which might direct experimental assays for functional investigations. Software tools and web services are often used for carrying of the bioinformatics analysis. After analysis of DNA, RNA and protein sequences, it is important to understand how they are connected by protein to genome mapping.

Identification of Genes in a Genomic DNA Sequence

Prediction of Protein-coding Genes

Archaeal and bacterial genes typically comprise uninterrupted stretches of DNA between a start codon (usually ATG, but in a minority of genes, GTG, TTG, or CTG) and a stop codon (TAA, TGA, or TAG; alternative genetic codes of certain bacteria, such as mycoplasmas, have only two stop codons). Rare exceptions to this rule involve important but rare mechanisms, such as programmed frameshifts. There seem to be no strict limits on the length of the genes. Indeed, the gene *rpmJ* encoding the ribosomal protein L36 is only 111 bp long in most bacteria, whereas the gene for *B. subtilis* polyketide synthase PksK is 13,343 bp long. In practice, mRNAs shorter than 30 codons are poorly translated, so protein-coding genes in prokaryotes are usually at least 100 bases in length. In prokaryotic genome-sequencing projects, open reading frames (ORFs) shorter than 100 bases are rarely taken into consideration, which does not seem to result in substantial underprediction. In contrast, in multicellular eukaryotes, most genes are interrupted by introns. The mean length of an exon is ~50 codons, but some exons are much shorter; many of the introns are extremely long, resulting in genes occupying up to several megabases of genomic DNA. This makes prediction of eukaryotic genes a far more complex (and still unsolved) problem than prediction of prokaryotic genes.

Prokaryotes

For most common purposes, a prokaryotic gene can be defined simply as the longest ORF for a given region of DNA. Translation of a DNA sequence in all six reading frames is a straightforward task, which can be performed on line using.

Of course, this approach is oversimplified and may result in a certain number of incorrect gene predictions, although the error rate is rather low. Firstly, DNA sequencing errors may result in incorrectly assigned or missed start and/or stop codons, because of which a gene might be truncated, overextended, or missed altogether. Secondly, on rare occasions, among two overlapping ORFs (on the same or the opposite DNA strand), the shorter one might be the real gene. The existence of a long "shadow" ORF opposite a protein-coding sequence is more likely than in a random sequence because of the statistical properties of the coding regions. Indeed, consider the simple case where the first base in a codon is a purine and the third base is a pyrimidine (the RNY codon pattern). Obviously, the mirror frame in the complementary strand would follow the same pattern, resulting in a deficit of stop codons. Figure below shows the ORFs of at least 100 bp located in a 10-kb fragment of the *E. coli* genome (from 3435250 to 3445250) that encodes potassium transport protein TrkA, mechanosensitive channel MscL, transcriptional regulator YhdM, RNA polymerase alpha subunit RpoA, preprotein translocase subunit SecY, and ribosomal proteins RplQ (L17), RpsD (S4), RpsK (S11), RpsM (S13), RpmJ (L36), RplO (L15), RpmD (L30), RpsE (S5), RplR (L18), RplF (L6), RpsH (S8), RpsN (S14), RplE (L5), and RplX (L24). Although the two ORFs in frame +1 (top line, on the right) are longer (207 aa and 185 aa) than the ORFs in frame −3 (bottom line, 117 aa, 177 aa, 130 aa, and 101 aa), it is the latter that encode real proteins, namely the ribosomal proteins RplR, RplF, RpsH, and RpsN.

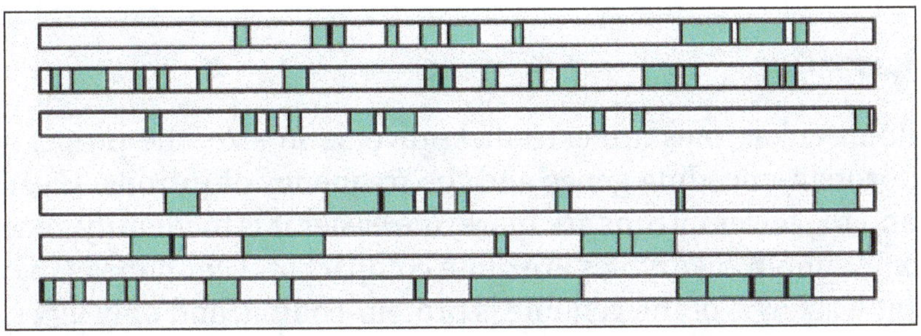

Open reading frames of ≥100 bp encoded on a 10-kb fragment of the *Escherichia coli* K12 genome from 3435250 to 3445250.

Because of these complications, it is always desirable to have some additional evidence that a particular ORF actually encodes a protein. Such evidence comes along many different lines and can be obtained using various methods, e.g. the following ones:

- The ORF in question encodes a protein that is similar to previously described ones (search the protein database for homologs of the given sequence).
- The ORF has a typical GC content, codon frequency, or oligonucleotide composition (calculate the codon bias and/or other statistical features of the sequence, compare to those for known protein-coding genes from the same organism).
- The ORF is preceded by a typical ribosome-binding site (search for a Shine-Dalgarno sequence in front of the predicted coding sequence).
- The ORF is preceded by a typical promoter (if consensus promoter sequences for the given organism are known, check for the presence of a similar upstream region).

The most reliable of these approaches is a database search for homologs. In several useful tools, DNA translation is seamlessly bound to the database searches. In the ORF finder, for example, the user can submit the translated sequence for a BLASTP or TBLASTN) search against the NCBI sequence databases. In addition, there is an opportunity to compare the translated sequence to the COG database. A largely similar Analysis and Annotation Tool, developed by Xiaoqiu Huang at Michigan Tech, also compares the translated protein sequences to nr and SWISS-PROT; in addition, it checks them against two cDNA databases, the dbEST at the NCBI and Human Gene Index at TIGR.

Other methods take advantage of the statistical properties of the coding sequences. For organisms with highly biased GC content, for example, the third position in each codon has a highly biased (very high or very low) frequency of G and C. FramePlot, a program that exploits this skew for gene recognition, is available at the Japanese Institute of Infectious Diseases and at the TIGR web site . The most useful and popular gene prediction programs, such as GeneMark and Glimmer, build Markov models of the known coding regions for the given organism and then employ them to estimate the coding potential of uncharacterized ORFs.

Inferring genes based on the coding potential and on the similarity of the encoded protein sequences to those of other proteins represent the intrinsic and extrinsic approaches to gene prediction, which ideally should be combined. Two programs that implement such a combination, developed specifically for analysis of prokaryotic genomes, are ORPHEUS and CRITICA.

Unicellular Eukaryotes

Genomes of unicellular eukaryotes are extremely diverse in size, the proportion of the genome that is occupied by protein-encoding genes and the frequency of introns. Clearly, the smaller the intergenic regions and the fewer introns are there, the easier it is to identify genes. Fortunately, genomes of at least some simple eukaryotes are quite compact and contain very few introns. Thus, in yeast *S. cerevisiae*, at least 67% of the genome is protein-coding, and only 233 genes (less than 4% of the total) appear to have introns. Although these include some biologically important and extensively studied genes, e.g. those for aminopeptidase APE2, ubiquitin-protein ligase UBC8, subunit 1 of the mitochondrial cytochrome oxidase COX1, and many ribosomal proteins, introns comprise less than 1% of the yeast genome. The tiny genome of the intracellular eukaryotic parasite *Encephalitozoon cuniculi* appears to contain introns in only 12 genes and is practically prokaryote-like in terms of the “wall-to-wall” gene arrangement. Malaria parasite *Plasmodium falciparum* is a more complex case, with ~43% of the genes located on chromosome 2 containing one or more introns. Protists with larger genomes often have fairly high intron density. In the slime mold *Physarum polycephalum*, for example, the average gene has 3.7 introns. Given that the average exon size in this organism (165 ± 85 bp) is comparable to the length of an average intron (138 ± 103 bp), homology-based prediction of genes becomes increasingly complicated.

Because of this genome diversity, there is no single way to efficiently predict protein-coding genes in different unicellular eukaryotes. For some of them, such as yeast, gene prediction can be done by using more or less the same approaches that are routinely employed in prokaryotic genome analysis. For those with intron-rich genomes, the gene model has to include information on the intron splice sites, which can be gained from a comparison of the genomic sequence against a set of ESTs from the same organism. This necessitates creating a comprehensive library of ESTs that have to be sequenced in a separate project. Such dual EST/genomic sequencing projects are currently under way for several unicellular eukaryotes.

Multicellular Eukaryotes

In most multicellular eukaryotes, gene organization is so complex that gene identification poses a major problem. Indeed, eukaryotic genes are often separated by large intergenic regions, and the genes themselves contain numerous introns, many of them long. Figure shows a typical distribution of exons and introns in a human gene, the X chromosome-located gene encoding iduronate 2-sulfatase (IDS_HUMAN), a lysosomal enzyme responsible for removing sulfate groups from heparan sulfate and dermatan sulfate. Mutations causing iduronate sulfatase deficiency result in the lysosomal accumulation of these glycosaminoglycans, clinically known as Hunter's syndrome or type II mucopolysaccharidosis (OMIM entry 309900). A number of clinical cases have been shown to result from aberrant alternative splicing of this gene's mRNA, which emphasizes the importance of reliable prediction of gene structure.

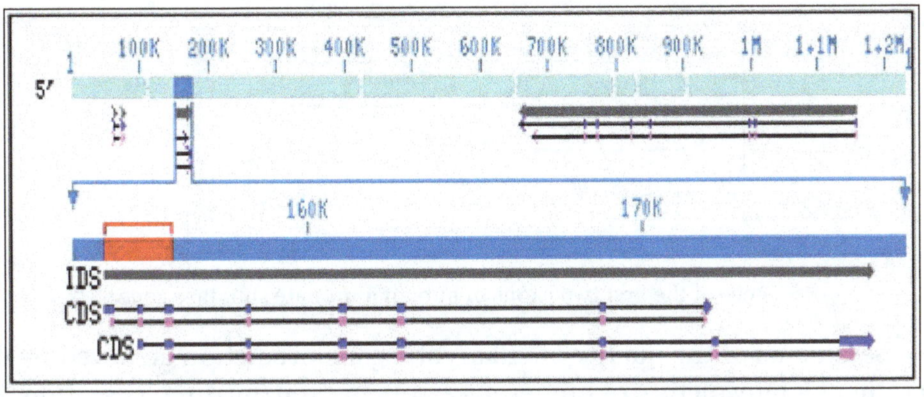

Organization of the human iduronate 2-sulfatase gene.

Obviously, the coding regions compose only a minor portion of the gene. In this case, positions of the exons could be unequivocally determined by mapping the cDNA sequence (i.e. iduronate sulfatase mRNA) back to the chromosomal DNA. Because of the clinical phenotype of the mutations in the iduronate sulfatase gene, we already know the "correct" mRNA sequence and can identify various alternatively spliced variants as mutations. However, for many, perhaps the majority of the human genes, multiple alternative forms are part of the regular expression pattern, and correct gene prediction ideally should identify all of these forms, which immensely complicates the task.

Ideally, gene prediction should identify all exons and introns, including those in the 5′-untranslated region (5′-UTR) and the 3′-UTR of the mRNA, in order to precisely reconstruct the predominant mRNA species. For practical purposes, however, it is useful to assemble at least the coding exons correctly because this allows one to deduce the protein sequence.

Correct identification of the exon boundaries relies on the recognition of the splice sites, which is facilitated by the fact that the great majority of splice sites conform to consensus sequences that include two nearly invariant dinucleotides at the ends of each intron, a GT at the 5′ end and an AG at the 3′ end. Non-canonical splice signals are rare and come in several variants. In the 5′ splice sites, the GC dinucleotide is sometimes found instead of GT. The second class of exceptions to the splice site consensus includes so-called "AT-AC" introns that have the highly conserved /(A,G) TATCCT(C,T) sequence at their 5′ sites. There are additional variants of non-canonical splice signals, which further complicate prediction of the gene structure.

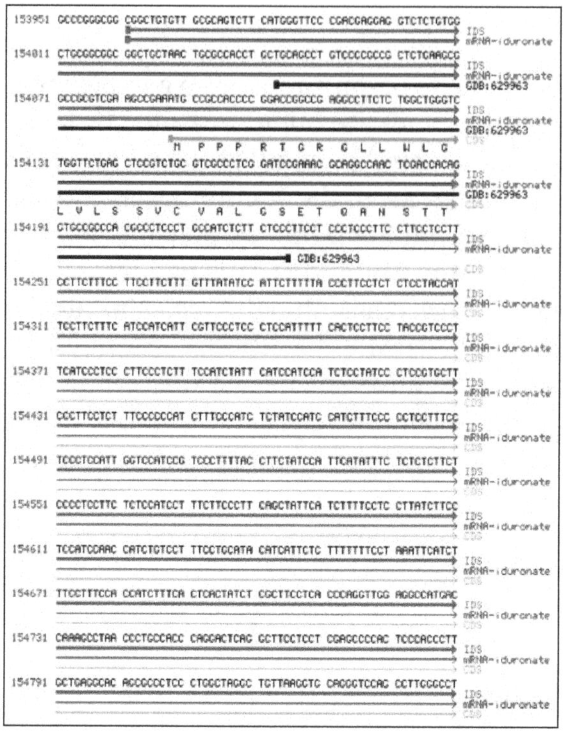

Sequence of the first two exons of human iduronate sulfatase gene.

The available assessments of the quality of eukaryotic gene prediction achieved by different programs show a rather gloomy picture of numerous errors in exon/intron recognition. Even the best tools correctly predict only ~40% of the genes. The most serious errors come from genes with long introns, which may be predicted as intragenic sequences, resulting in erroneous gene fission, and pairs of genes with short intergenic regions, which may be predicted as introns, resulting in false gene fusion. Nevertheless, most of the popular gene prediction programs show reasonable performance in predicting the coding regions in the sense that, even if a small exon is missed or overpredicted, the majority of exons are identified correctly.

Another important parameter that can affect ORF prediction is the fraction of sequencing errors in the analyzed sequence. Indeed, including frameshift corrections was found to substantially improve the overall quality of gene prediction. Several algorithms were described that could detect frameshift errors based on the statistical properties of coding sequences. On the other hand, error correction techniques should be used with caution because eukaryotic genomes contain numerous pseudogenes, and non-critical frameshift correction runs the risk of wrongly "rescuing" pseudogenes. The problem of discriminating between pseudogenes and frameshift errors is actually quite complex and will likely be solved only through whole-genome alignments of different species or, in certain cases, by direct experimentation, e.g., expression of the gene(s) in question.

Algorithms and Software Tools for Gene Identification

Recognizing genes in the DNA sequences remains one of the most pressing problems in genome analysis. Several different approaches to gene prediction have been developed, and there are several popular programs that are most commonly used for this task. Some of these tools perform gene prediction *ab initio*, relying only on the statistical parameters in the DNA sequence for gene identification. In contrast, homology-based methods rely primarily on identifying homologous sequences in other genomes and/or in public databases using BLAST or Smith-Waterman algorithms. Many of the commonly used methods combine these two approaches.

The absence of introns and relatively high gene density in most genomes of prokaryotes and some unicellular eukaryotes provides for effective use of sequence similarity searches as the first step in genome annotation. Genes identified by homology can be used as the training set for one of the statistical methods for gene recognition, and the resulting statistical model can then be used for analyzing the remaining parts of the genome. In most eukaryotes, the abundance of introns and long intergenic regions makes it difficult to use homology-based methods as the first step unless, of course, one can rely on synteny between several closely related genomes (e.g. human, mouse, and rat). As a result, gene prediction for genome sequences of multicellular eukaryotes usually starts with *ab initio* methods, followed by similarity searches with the initial exon assemblies.

Therefore, it is advisable to use at least two different programs for gene prediction in a new DNA sequence, especially if it comes from a eukaryote or a poorly characterized prokaryote. A comparison of predictions generated by different programs reveals the cases where a given program performs the best and helps in achieving consistent quality of gene prediction. Such a comparison can be performed, for example, using the TIGR Combiner program which employs a voting scheme to combine predictions of different gene-finding programs, such as GeneMark, GlimmerM, GRAIL, GenScan, and Fgenes.

GeneMark

GeneMark, mirrored at the EBI web site was developed by Mark Borodovsky and James McIninch in 1993. GeneMark was the first tool for finding prokaryotic genes that employed a non-homogeneous Markov model to classify DNA regions into protein-coding, non-coding, and non-coding but complementary to coding. It has been shown previously that, by multivariate codon usage analysis, the *E. coli* genes could be classified into so-called typical, highly typical, and atypical gene sets, with the latter two groups apparently corresponding to highly expressed genes and horizontally transferred genes. Accordingly, more than one Markov model was required to adequately describe different groups of genes in the same genome.

Like other gene prediction programs, GeneMark relies on organism-specific recognition parameters to partition the DNA sequence into coding and non-coding regions and thus requires a sufficiently large training set of known genes from a given organism for best performance. The program has been repeatedly updated and modified and now exists in separate variants for gene prediction in prokaryotic, eukaryotic, and viral DNA sequences.

Glimmer

Gene Locator and Interpolated Markov Modeler (GLIMMER, developed by Steven Salzberg and colleagues at Johns Hopkins University and TIGR, is a system for finding genes in prokaryotic genomes. To identify coding regions and distinguish them from noncoding DNA, Glimmer uses interpolated Markov models, i.e. series of Markov models with the order of the model increasing at each step and the predictive power of each model separately evaluated. Like GeneMark, Glimmer requires a training set, which is usually selected among known genes, genes coding for proteins with strong database hits, and/or simply long ORFs. Glimmer is used as the primary gene finder tool at TIGR, where it has been applied to the annotation of numerous microbial genomes.

Recently, Salzberg and coworkers developed GlimmerM, a modified version of Glimmer specifically designed for gene recognition in small eukaryotic genomes, such as the malaria parasite *Plasmodium falciparum*.

Grail

Gene Recognition and Assembly Internet Link, developed by Ed Uberbacher and coworkers at the Oak Ridge National Laboratory, is a tool that identifies exons, polyA sites, promoters, CpG islands, repetitive elements, and frameshift errors in DNA sequences by comparing them to a database of known human and mouse sequence elements. Exon and repetitive element prediction is also available for *Arabidopsis* and *Drosophila* sequences.

Grail has been recently incorporated into the Oak Ridge genome analysis pipeline (, which provides a unified web interface to a number of convenient analysis tools. For prokaryotes, it offers gene prediction using Glimmer and Generation programs, followed by BLASTP searches of predicted ORFs against SWISS-PROT and NR databases and a HMMer search against Pfam. There is also an option of BLASTN search of the submitted DNA sequence against a variety of nucleotide sequence databases.

For human and mouse sequences, the Oak Ridge pipeline offers gene prediction using GrailEXP and GenScan, also followed by BLASTP searches of predicted ORFs against SWISS-PROT and NR databases and a HMMer search against Pfam. Again, the user can perform BLASTN search of the submitted DNA sequence against a variety of nucleotide sequence databases, as well as search for CpG islands, repeat fragments, tRNAs and BAC-end pairs. the possibility to directly compare gene predictions made by two different programs is a valuable feature, which is available at the Oak Ridge web site.

GenScan

GenScan as developed by Chris Burge and Samuel Karlin at Stanford University and is currently hosted in the Burge laboratory at the MIT Department of Biology. This program uses a complex probabilistic model of the gene structure that is based on actual biological information about the properties of transcriptional, translational, and splicing signals. In addition, it utilizes various statistical properties of coding and noncoding regions. To account for the heterogeneity of the human genome that affects gene structure and gene density, GenScan derives different sets of gene models for genome regions with different GC content. Its high speed and accuracy make GenScan the method of choice for the initial analysis of large (in the megabase range) stretches of eukaryotic genomic DNA. GenScan has being used as the principal tool for gene prediction in the International Human Genome Project.

GeneBuilder

GeneBuilder performs *ab initio* gene prediction using numerous parameters, such as GC content, dicodon frequencies, splicing site data, CpG islands, repetitive elements, and others. It also utilizes a unique approach that is based on evaluating relative frequencies of synonymous and nonsynonymous substitutions to identify likely coding sequences. In addition, it performs BLASTsearches of predicted genes against protein and EST databases, which helps to refine the boundaries of predicted exons using the BLAST hits as guides. The program allows the user to change certain parameters, which permits interactive gene structure prediction. As a result, GeneBuilder is sometimes able to predict the gene structure with a good accuracy, even when the similarity of the predicted ORF to a homologous protein sequence is low.

Splice Site Prediction

Programs for predicting intron splice sites, which are commonly used as subroutines in the gene prediction tools, can also be used as stand-alone programs to verify positions of splice sites or predict alternative splicing sites. Such programs can be particularly useful for predicting non-coding exons, which are commonly missed in gene prediction studies.

Recognition of the splice sites by these programs usually relies on statistical properties of exons and introns and on the consensus sequences of splicing signals. A detailed study of the performance of one such program, SpliceView, showed that, although the fraction of missed splicing signals was relatively low (~5%), the false-positive rate was quite high (typically, one potential splicing signal per 150–250 bases). One should note, however, that such false-positive signals might correspond to rare alternative splice forms or cryptic splice sites (splice sites that are not active in normal genes and become activated as a result of mutations in major splicing signals).

Combining Various Gene Prediction Tools

While the first step of gene identification in long genomic sequences utilizes *ab initio*programs that can rapidly and with reasonable accuracy predict multiple genes, the next step validates these predictions through similarity searches. Predicted genes are compared to nucleotide sequence databases, including EST databases and protein sequences encoded by these predicted genes are compared to protein sequence databases. These data are then combined with the information about repetitive elements, CpG islands, and transcription factor binding sites and used for further refinement of gene structure. Thus, homology information is ultimately incorporated into every gene prediction pipeline. There are, however, several programs that primarily rely on similarity search for gene prediction. Although differing in details, they all search for the best alignment of the given piece of DNA to the homologous nucleotide or protein sequences in the database.

Principles of Sequence Similarity Searches

As discussed in the previous section, initial characterization of any new DNA or protein sequence starts with a database search aimed at finding out whether homologs of this gene (protein) are already available, and if they are, what is known about them? Clearly, looking for exactly the same sequence is quite straightforward. One can just take the first letter of the query sequence, search for its first occurrence in the database, and then check if the second letter of the query is the same in the subject. If it is indeed the same, the program could check the third letter, then the fourth, and continue this comparison to the end of the query. If the second letter in the subject is different from the second letter in the query, the program should search for another occurrence of the first letter, and so on. This will identify all the sequences in the database that are identical to the query sequence (or include it). Of course, this approach is primitive computation-wise, and there are sophisticated algorithms for text matching that do it much more efficiently.

```
Query: 1 MK  → Query: 1 MKV  → Query: 1 MKVR  → Query: 1 MKVRA
Sbjct: 1 MK    Sbjct: 1 MKV    Sbjct: 1 MKVR    Sbjct: 1 MKVRA

... → Query: 1 MKVRASVKKLCRNCKIVKRDGVIRVICSAEPKHKQRQG
      Sbjct: 1 MKVRASVKKLCRNCKIVKRDGVIRVICSAEPKHKQRQG
```

Note that, in the example above, we looked only for sequences that ***exactly match***the query. The algorithm would not even find a sequence that is identical to the query with the exception of the first letter. To find such sequences, the same analysis should be conducted with the fragments starting from the second letter of the original query, then from the third one, and so on.

```
Query1: 1   KVRASVKKLCRNCKIVKRDGVIRVICSAEPKHKQRQG
Query2: 1    VRASVKKLCRNCKIVKRDGVIRVICSAEPKHKQRQG
Query3: 1     RASVKKLCRNCKIVKRDGVIRVICSAEPKHKQRQG
Query4: 1      ASVKKLCRNCKIVKRDGVIRVICSAEPKHKQRQG
```

Such search quickly becomes time-consuming, and we are still dealing only with identical sequences. Finding close relatives would introduce additional conceptual and technical problems. Let us assume that sequences that are 99% identical are definitely homologous. What should one select as the threshold to consider sequences not to be homologous: 50% identity, 33%, or perhaps 25%? These are legitimate questions that need to be answered before one goes any further. The example of two lysozymes shows that sequences with as low as 8% identity may belong to orthologous proteins and perform the same function.

As a matter of fact, when comparing nucleic acid sequences, there is very little one could do. All the four nucleotides, A, T, C, and G, are found in the database with approximately the same frequencies and have roughly the same probability of mutating one into another. As a result, DNA-DNA comparisons are largely based on straightforward text matching, which makes them fairly slow and not particularly sensitive, although a variety of heuristics have been developed to overcome this.

Amino acid sequence comparisons have several distinct advantages over nucleotide sequence comparisons, which, at least potentially, lead to a much greater sensitivity. Firstly, because there are 20 amino acids but only four bases, an amino acid match carries with it >4 bits of information as opposed to only two bits for a nucleotide match. Thus, statistical significance can be ascertained for much shorter sequences in protein comparisons than in nucleotide comparisons. Secondly, because of the redundancy of the genetic code, nearly one-third of the bases in coding regions are under a weak (if any) selective pressure and represent noise, which adversely affects the sensitivity of the searches. Thirdly, nucleotide sequence databases are much larger than protein databases because of the vast amounts of non-coding sequences coming out of eukaryotic genome projects, and this further lowers the search sensitivity. Finally, and probably most importantly, unlike in nucleotide sequence, the likelihoods of different amino acid substitutions occurring during evolution are substantially different, and taking this into account greatly improves the performance of database search methods. Given all these advantages, comparisons of any coding sequences are typically carried out at the level of protein sequences; even when the goal is to produce a DNA-DNA alignment (e.g. for analysis of substitutions in silent codon positions), it is usually first done with protein sequences, which are then replaced by the corresponding coding sequences. Direct nucleotide sequence comparison is indispensable only when non-coding regions are analyzed.

Substitution Scores and Substitution Matrices

The fact that each of the 20 standard protein amino acids has its own unique properties means that the likelihood of the substitution of each particular residue for another residue during evolution should be different. Generally, the more similar the physico-chemical properties of two residues, the greater the chance that the substitution will not have an adverse effect on the protein's function and, accordingly, on the organism's fitness. Hence, in sequence comparisons, such a substitution should be penalized less than a replacement of amino acid residue with one that has dramatically different properties. This is, of course, an oversimplification, because the effect of a substitution depends on the structural and functional environment where it occurs. For example, a cysteine-to-valine substitution in the catalytic site of an enzyme will certainly abolish the activity and, on many occasions, will have a drastic effect on the organism's fitness. In contrast, the same substitution within a β-strand may have little or no effect. Unfortunately, in general, we do not have *a priori*knowledge of the location of a particular residue in the protein structure, and even with such knowledge, incorporating it in a database search algorithm is an extremely complex task. Thus, a generalized measure of the likelihood of amino acid substitutions is required so that each substitution is given an appropriate score (weight) to be used in sequence comparisons. The score for a substitution between amino acids *i* and *j* always can be expressed by the following intuitively plausible formula, which shows how likely a particular substitution is given the frequencies of each the two residues in the analyzed database:

$$S_{IJ} = kln(q_{ij} / p_i p_j)$$

where *k* is a coefficient, q_{ij} is the observed frequency of the given substitution, and p_i, p_j are the background frequencies of the respective residues. Obviously, here the product $p_i p_j$ is the expected frequency of the substitution and, if $q_{ij} = p_i p_j$ ($S_{ij} = 0$), the substitution occurs just as often as expected. The scores used in practice are scaled such that the expected score for aligning a random pair of amino acid sequences is negative.

There are two fundamentally different ways to come up with a substitution score matrix, i.e. a triangular table containing 210 numerical score values for each pair of amino acids, including identities (diagonal elements of the matrix; as shown in figure below). As in many other situations in computational biology, the first approach works *ab initio*, whereas the second one is empirical. One *ab initio* approach calculates the score as the number of nucleotide substitutions that are required to transform a codon for one amino acid in a pair into a codon for the other. In this case, the matrix is obviously unique (as long as alternative genetic codes are not considered) and contains only four values, 0, 1, 2, or 3. Accordingly, this is a very coarse grain matrix that is unlikely to work well. The other *ab initio* approach assigns scores on the basis of similarities and differences in the physico-chemical properties of amino acids. Under this approach, the number of possible matrices is infinite, and they may have as fine a granularity as desirable, but a degree of arbitrariness is inevitable because our understanding of protein physics is insufficient to make informed decisions on what set of properties "correctly" reflects the relationships between amino acids.

	A	R	N	D	C	Q	E	G	H	I	L	K	M	F	P	S	T	W	Y	V	B	Z	X
A	6																						
R	-7	8																					
N	-4	-6	8																				
D	-3	-10	2	8																			
C	-6	-8	-11	-14	10																		
Q	-4	-2	-3	-2	-14	8																	
E	-2	-9	-2	2	-14	1	8																
G	-2	-9	-3	-3	-9	-7	-4	6															
H	-7	-2	0	-4	-7	1	-5	-9	9														
I	-5	-5	-5	-7	-6	-8	-5	-11	-9	8													
L	-6	-8	-7	-12	-15	-5	-9	-10	-6	-1	7												
K	-7	0	-1	-4	-14	-3	-4	-7	-6	-6	-8	7											
M	-5	-4	-9	-11	-13	-4	-7	-8	-10	-1	1	-2	11										
F	-8	-9	-9	-15	-13	-13	-14	-9	-6	-2	-3	-14	-4	9									
P	-2	-4	-6	-8	-8	-3	-5	-6	-4	-8	-7	-6	-8	-10	8								
S	0	-3	0	-4	-3	-5	-4	-2	-6	-7	-8	-4	-5	-6	-2	6							
T	-1	-6	-2	-5	-8	-5	-6	-6	-7	-2	-7	-3	-4	-9	-4	0	7						
W	-13	-2	-8	-15	-15	-13	-17	-15	-7	-14	-6	-12	-13	-4	-14	-5	-13	13					
Y	-8	-10	-4	-11	-4	-12	-8	-14	-3	-6	-7	-9	-11	2	-13	-7	-6	-5	10				
V	-2	-8	-8	-8	-6	-7	-6	-5	-6	2	-2	-9	-1	-8	-6	-6	-3	-15	-7	7			
B	-3	-7	6	6	-12	-3	1	-3	-1	-6	-9	-2	-10	-10	-7	-1	-3	-10	-6	-8	6		
Z	-3	-4	-3	1	-14	6	6	-5	-1	-6	-7	-4	-5	-13	-4	-5	-6	-14	-9	-6	0	6	
X	-3	-6	-3	-5	-9	-5	-5	-5	-5	-5	-6	-5	-5	-8	-5	-3	-4	-11	-7	-5	-5	-5	-5

The PAM30 substitution matrix.

	A	R	N	D	C	Q	E	G	H	I	L	K	M	F	P	S	T	W	Y	V	X
R	-1	5																			
N	-2	0	6																		
D	-2	-2	1	6																	
C	0	-3	-3	-3	9																
Q	-1	1	0	0	-3	5															
E	-1	0	0	2	-4	2	5														
G	0	-2	0	-1	-3	-2	-2	6													
H	-2	0	1	-1	-3	0	0	-2	8												
I	-1	-3	-3	-3	-1	-3	-3	-4	-3	4											
L	-1	-2	-3	-4	-1	-2	-3	-4	-3	2	4										
K	-1	2	0	-1	-3	1	1	-2	-1	-3	-2	5									
M	-1	-1	-2	-3	-1	0	-2	-3	-2	1	2	-1	5								
F	-2	-3	-3	-3	-2	-3	-3	-3	-1	0	0	-3	0	6							
P	-1	-2	-2	-1	-3	-1	-1	-2	-2	-3	-3	-1	-2	-4	7						
S	1	-1	1	0	-1	0	0	0	-1	-2	-2	0	-1	-2	-1	4					
T	0	-1	0	-1	-1	-1	-1	-2	-2	-1	-1	-1	-1	-2	-1	1	5				
W	-3	-3	-4	-4	-2	-2	-3	-2	-2	-3	-2	-3	-1	1	-4	-3	-2	11			
Y	-2	-2	-2	-3	-2	-1	-2	-3	2	-1	-1	-2	-1	3	-3	-2	-2	2	7		
V	0	-3	-3	-3	-1	-2	-2	-3	-3	3	1	-2	1	-1	-2	-2	0	-3	-1	4	
X	0	-1	-1	-1	-2	-1	-1	-1	-1	-1	-1	-1	-1	-1	-2	0	0	-2	-1	-1	-1

The BLOSUM 62 substitution matrix.

Empirical approaches, which historically came first, attempt to derive the characteristic frequencies of different amino acid substitutions from actual alignments of homologous protein families. In other words, these approaches strive to determine the actual likelihood of each substitution occurring during evolution. Obviously, the outcome of such efforts critically depends on the quantity and quality of the available alignments, and even now, any alignment database is far from being complete or perfectly correct. Furthermore, simple counting of different types of substitutions will not suffice if alignments of distantly related proteins are included because, in many cases, multiple substitutions might have occurred in the same position. Ideally, one should construct the phylogenetic tree for each family, infer the ancestral sequence for each internal node, and then count the substitutions exactly. This is not practicable in most cases, and various shortcuts need to be taken.

Several solutions to these problems have been proposed, each resulting in a different set of substitution scores. The first substitution matrix, constructed by Dayhoff and Eck in 1968, was based on an alignment of closely related proteins, so that the ancestral sequence could be deduced and all the amino acid replacements could be considered occurring just once. This model was then extrapolated to account for more distant relationships, which resulted in the PAM series of substitution matrices. PAM (Accepted Point Mutation) is a unit of evolutionary divergence of protein sequences, corresponding to one amino acid change per 100 residues. Thus, for example, the PAM30 matrix is supposed to apply to proteins that differ, on average, by 0.3 changes per aligned residue, whereas PAM250 should reflect evolution of sequences with an average of 2.5 substitutions per position. Accordingly, the former matrix should be employed for constructing alignments of closely related sequences, whereas the latter is useful in database searches aimed at detection of distant relationships. Using an approach similar to that of Dayhoff, combined with rapid algorithms for protein sequence clustering and alignment, Jones, Taylor, and Thornton produced the series of the so-called JTT matrices, which are essentially an update of the PAMs.

The PAM and JTT matrices, however, have obvious limitations because of the fact that they have been derived from alignments of closely related sequences and extrapolated to distantly related ones. This extrapolation may not be fully valid because the underlying evolutionary model might not be adequate, and the trends that determine sequence divergence of closely related sequences might not apply to the evolution at larger distances.

In 1992, Steven and Jorja Henikoff developed a different series of substitution matrices using conserved ungapped alignments of related proteins from the BLOCKS database. The use of these alignments offered three important advantages over the alignments used for constructing the PAM matrices. First, the BLOCKS collection obviously included a much larger number and, more importantly, a much greater diversity of protein families than the collection that was available to Dayhoff and coworkers in the 1970's. Second, coming from rather distantly related proteins, BLOCKS alignments better reflected the amino acid changes that occur over large phylogenetic distances and thus produced substitution scores that represented sequence divergence in distant homologs directly, rather than through extrapolation. Third, in these distantly related proteins, BLOCKS included only the most confidently aligned regions, which are likely to best represent the prevailing evolutionary trends. These substitution matrices, named the BLOSUM (= BLOcks SUbstitution Matrix) series, were tailored to particular evolutionary distances by ignoring the sequences that had more than a certain percent identity. In the BLOSUM62 matrix, for example, the substitution scores were derived from the alignments of sequences that had no more than 62% identity; the substitution scores of the BLOSUM45 matrix were calculated from the alignments that contained sequences with no more than 45% identity. Accordingly, BLOSUM matrices with high numbers, such as BLOSUM80, are best suited for comparisons of closely related sequences (it is also advisable to use BLOSUM80 for database searches with short sequences,, whereas low-number BLOSUM matrices, such as BLOSUM45, are better for distant relationships. In addition to the general-purpose PAM, JTT, and BLOSUM series, some specialized substitution matrices were developed, for example, for integral membrane proteins, but they never achieved comparable recognition.

Several early studies found the PAM matrices based on empirical data consistently resulted in greater search sensitivity than any of the *ab initio* matrices. An extensive empirical comparison showed that: (i) BLOSUM matrices consistently outperformed PAMs in BLAST searches, and (ii) on average, BLOSUM62 performed best in the series; this matrix is currently used as the default in most sequence database searches. It is remarkable that so far, throughout the 30-plus-year history of amino acid substitution matrices, empirical matrices have consistently outperformed those based on theory, either physico-chemical or evolutionary. This is not to say, of course, that theory is powerless in this field, but to point out that we currently do not have a truly adequate theory to describe protein evolution. Clearly, the last word has not been said on amino acid substitution matrices, and one can expect that eventually the BLOSUM series will be replaced by new matrices based on greater amounts of higher quality alignment data and more realistic evolutionary models. A recently reported maximum-likelihood model for substitution frequency estimation has already been claimed to describe individual protein families better than the Dayhoff and JTT models. It remains to be seen how this and other new matrices perform in large-scale computational experiments on real databases. AAindex lists 66 different substitution matrices, both *ab initio* and empirical, and there is no doubt that this list will continue to grow.

Statistics of Protein Sequence Comparison

It is impossible to explain even the basic principles of statistical analysis of sequence similarities without invoking some mathematics. To introduce these concepts in the least painful way, let us consider the same protein sequence as above and check how many times segments of this sequence of different lengths are found in the database (we chose fragments starting from the second position in the sequence because nearly every protein in the database starts with a methionine). Not unexpectedly, we find that the larger the fragment, the smaller the number of exact matches in the database.

```
Query: 1 MKVRASVKKLCRNCKIVKRDGVIRVICSAEPKHKQRQG 38
```

Table: Dependence of the number of exact database matches on the length of the query word.

Sequence	Occurrences in the database
KV	488,559
KVR	28,592
KVRA	2,077
KVRAS	124
KVRASV	23
KVRASVK	8
KVRASVKK	4
KVRASVKKL	1
KVRASVKKLC	1

Perhaps somewhat counterintuitively, a 9-mer is already unique. With the decrease in the number of database hits, the likelihood that these hits are biologically relevant, i.e. belong to homologs of the query protein, increases. Thus, 13 of the 23 occurrences of the string KVRASV and all 8 occurrences of the string KVRASVK are from RpsJ orthologs.

The number of occurrences of a given string in the database can be roughly estimated as follows. The probability of matching one amino acid residue is 1/20 (assuming equal frequencies of all 20 amino acids in the database; this not being the case, the probability is slightly greater). The probability of matching two residues in a row is then $(1/20)^2$, and the probability of matching *n* residues is $(1/20)^n$. Given that the protein database currently contains $N \sim 2 \times 10^8$ letters, one should expect a string of *n*letters to match approximately $N \times (1/20)^n$ times, which is fairly close to the numbers in table above.

Searching for perfect matches is the simplest and, in itself, obviously insufficient form of sequence database search, although, as we shall see below, it is important as one of the basic steps in currently used search algorithms.The goal of a search is finding homologs, which can have drastically different sequences such that, in distant homologs, only a small fraction of the amino acid residues are identical or even similar. Even in close homologs, a region of high similarity is usually flanked by dissimilar regions like in the following alignment of *E. coli* RpmJ with its ortholog from *Vibrio cholerae*:

```
E. coli RpmJ:  1 MKVRASVKKLCR---NCKIVKRDGVIRVICSAEPKHKQRQG
                 MKV +S+K       +C+IVKR G + VIC + P+ K  Q
Vibrio VC0879: 1 MKVLSSLKSAKNRHPDCQIVKRRGRLYVICKSNPRFKAVQR
```

In this example, the region of highest similarity is in the middle of the alignment, but including the less conserved regions on both sides improves the overall score (taking into account the special treatment of gaps. Further along the alignment, the similarity almost disappears so that inclusion of additional letters into the alignment would not increase the overall score or would even decrease it. Such fragments of the alignment of two sequences whose similarity score cannot be improved by adding or trimming any letters, are referred to as high-scoring segment pairs (HSPs). For this approach to work, the expectation of the score for random sequences must be negative, and the scoring matrices used in database searches are scaled accordingly.

So, instead of looking for perfect matches, sequence comparisons programs actually search for HSPs. Once a set of HSPs is found, different methods, such as Smith-Waterman, FAS TA, or BLAST, deal with them in different fashions.However, the principal issue that any database search method needs to address is identifying those HSPs that are unlikely to occur by chance and, by inference, are likely to belong to homologs and to be biologically relevant. This problem has been solved by Samuel Karlin and Stephen Altschul, who showed that maximal HSPscores follow the extreme value distribution. Accordingly, if the lengths of the query sequence (m) and the database (n) are sufficiently high, the expected number of HSPs with a score of at least S is given by the formula:

$$E = Kmn2^{-\lambda S}$$

Here, S is the so-called raw score calculated under a given scoring system, and K and λ are natural scaling parameters for the search space size and the scoring system, respectively. Normalizing the score according to the formula:

$$S' = (\lambda S - l11K) / ln2$$

gives the bit score, which has a standard unit accepted in information theory and computer science. Then,

$$E = mn2^{-S'}$$

and, since it can be shown that the number of random HSPs with score $\geq S'$ is described by Poisson distribution, the probability of finding at least one HSP with bit score $\geq S'$ is:

$$P = 1 - e^{-E}$$

Equation above links two commonly used measures of sequence similarity, the probability (P-value) and expectation (E-value). For example, if the score S is such that three HSPs with this score (or greater) are expected to be found by chance, the probability of finding at least one such HSP is ($1 - e^{-3}$), ~0.95. By definition, P-values vary from 0 to 1, whereas E-values can be much greater than 1. The BLAST programs report E-values, rather than P-values, because E-values of, for example, 5 and 10 are much easier to comprehend than P-values of 0.993 and 0.99995. However, for E < 0.01, P-value and E-value are nearly identical.

The product *mn* defines the *search space*, a critically important parameter of any database search. Equations $E = Kmn2^{-\lambda S}$ and $E = mn2^{-S'}$ codify the intuitively obvious notion that the larger the search space, the higher the expectation of finding an HSP with a score greater than any given value. There are two corollaries of this that might take some getting used to: (i) the same HSP may come out statistically significant in a small database and not significant in a large database; with the natural growth of the database, any given alignment becomes less and less significant (but by no means less important because of that) and (ii) the same HSP may be statistically significant in a small protein (used as a query) and not significant in a large protein.

Clearly, one can easily decrease the E-value and the P-value associated with the alignment of the given two sequences by lowering *n* in equation 4.2, i.e. by searching a smaller database. However, the resulting increase in significance is false, although such a trick can be useful for detecting initial hints of subtle relationships that should be subsequently verified using other approaches. The simple notion of E(P)-value is often misunderstood and interpreted as if these values applied just to a single pairwise comparison (i.e., if an E-value of 0.001 for an HSP with score *S* is reported, then, in a database of just a few thousand sequences, one expects to find a score >*S* by chance). It is critical to realize that the size of the search space is already factored in these E-values, and the reported value corresponds to the database size at the time of search (thus, it is certainly necessary to indicate, in all reports of sequence analysis, which database was searched, and desirably, also on what exact date).

Speaking more philosophically (or futuristically), one could imagine that, should the genomes of all species that inhabit this planet be sequenced, it would become almost impossible to demonstrate statistical significance for any but very close homologs in standard database searches. Thus, other approaches to homology detection are required that counter the problems created by database growth by taking advantage of the simultaneously increasing sequence diversity.

There has been rigorously proved to apply only to sequence alignments that do not contain gaps, whereas statistical theory for the more realistic gapped alignments remains an open problem. However, extensive computer simulations have shown that these alignments also follow the extreme value distribution to a high precision; therefore, at least for all practical purposes, the same statistical formalism is applicable.

Protein Sequence Complexity: Compositional Bias

The existence of a robust statistical theory of sequence comparison, in principle, should allow one to easily sort search results by statistical significance and accordingly assign a level of confidence to any homology identification. However, a major aspect of protein molecule organization substantially complicates database search interpretation and may lead to gross errors in sequence analysis. Many proteins, especially in eukaryotes, contain low (compositional) complexity regions, in which the distribution of amino acid residues is non-random, i.e. deviates from the standard statistical model. In other words, these regions typically have biased amino acid composition, e.g.

are rich in glycine or proline, or in acidic or basic amino acid residues. The ultimate form of low complexity is, of course, a homopolymer, such as a Q-linker. Other low-complexity sequences have a certain amino acid periodicity, sometimes subtle, such as, for example, in coiled-coil and other non-globular proteins (e.g. collagen or keratin).

The notion of compositional complexity was encapsulated in the SEG algorithm and the corresponding program, which partitions protein sequences into segments of low and high (normal) complexity. An important finding made by John Wootton is that low-complexity sequences correspond to non-globular portions of proteins. In other words, a certain minimal level of complexity is required for a sequence to fold into a globular structure. Low-complexity regions in proteins, although devoid of enzymatic activity, have important biological functions, most often promoting protein-protein interactions or cellular adhesion to various surfaces and to each other.

In a detailed empirical study, a set of parameters of the SEG program was identified that allowed reasonably accurate partitioning of a protein sequence into predicted globular and non-globular parts. The mastermind protein of Drosophila is a component of the Notch-dependent signaling pathway and plays an important role in the development of the nervous system of the fruit fly. In spite of this critical biological function, this protein consists mostly of stretches of only three amino acid residues, Gln, Asn, and Gly, and is predicted to have a predominantly non-globular structure. Recently discovered human homologs of mastermind are also involved in Notch-dependent transcriptional regulation and similarly appear to be almost entirely non-globular.

```
>gi|126721|sp|P21519|MAM_DROME NEUROGENIC PROTEIN
MASTERMIND
                                 1-13    MDAGGLPVFQSAS
qaaavaqqqqqqqqqhlnlqlhqqhlglh    14-85
lqqqqqlqlqqqqhnaqaqqqqiqvqqqqq
                  qqqqqqqqqqqh
                                 86-88   SPY
nanlgatggiagitggngaggptnpgavpt   89-121
                           apg
                                 122-195 DTMPTKRMPVVDRLRR
                                         RMENYRRRQTDCVPRY
                                         EQAFNTVCEQQNQETT
                                         VLQKRFLESKNKRAAK
                                         KTDKKLPDPS
        qqhqqqhqqqqhqqhqqhqq     196-217
                                 218-258 AQTMLAGQLQSSVHVQ
                                         QKFLKRPAEDVDNGPD
                                         SFEPPHKLP
nnnnnsnsnnnngnananngg ngsntgnnt  259-460
nnngnstnnnggsnnngsenltkfsveivq
qlefttspansqpqqistnvtvkaltntsv
ksepgvgggggggggnsgnnnnnggggg
gngnnnnnggdhhqqqqhqhqqqqqqqg
glgglgnngrgggpggmatgpggvagglgg
        mgmppnmmsaqqksalgnlanl
                                 461-465 VECKR
epdhdfpdlgsldkdggggqfpgfpdllgd   466-755
dnsenndtfkdlinnlqdfnpsfldgfdek
plldiktedgikveppnaqdlinslnvkse
gglghgfggfglgldnpgmkmrggnpgnqg
gfpngpnggtggapnaggnggnsgnlmseh
plaaqtlkqmaeqhqhknamggmggfprpp
hgmnpqqqqqqqqqqqqqaqqqhgqmmgq
gqpgryndygggfpndfglgpngpqqqqq
```

Sequence of the Drosophila mastermind protein: partitioning into predicted non-globular (left column) and globular (right column) regions.

Low-complexity regions represent a major problem for database searches. Since the λ parameter of equation $E = Kmn2^{-\lambda S}$ is calculated for the entire database, Karlin-Altschul statistics breaks down when the composition of the query or a database sequence or both significantly deviates from the average composition of the database. The result is that low-complexity regions with similar composition (e.g. acidic or basic) often produce "statistically significant" alignments that have nothing to with homology and are completely irrelevant. The SEG program can be used to overcome this problem in a somewhat crude manner: the query sequence, the database, or both can be partitioned into normal complexity and low-complexity regions, and the latter are masked (i.e. amino acid symbols are replaced with the corresponding number of X's). For the purpose of a database search, such filtering is usually done using short windows so that only the segments with a strongly compositional bias are masked. Low-complexity filtering has been indispensable for making database search methods, in particular BLAST, into reliable tools. Without masking low-complexity regions, false results would have been produced for a substantial fraction of proteins, especially eukaryotic ones (an early estimate held that low-complexity regions comprise ~15% of the protein sequences in the SWISS-PROT database). These false results would have badly polluted any large-scale database search, and the respective proteins would have been refractory to any meaningful sequence analysis. For these reasons, for several years, SEG filtering had been used as the default for BLAST searches to mask low-complexity segments in the query sequence. However, this procedure is not without its drawbacks. Not all low-complexity sequences are captured, and false-positives still occur in database searches. The opposite problem also hampers database searches for some proteins: when short low-complexity sequences are parts of conserved regions, statistical significance of an alignment may be underestimated, sometimes grossly.

In a recent work of Alejandro Schäffer and colleagues, a different, less arbitrary approach for dealing with compositionally biased sequences was introduced. This method, called composition-based statistics, recalculates the λ parameter and, accordingly, the E values for each query and each database sequence, thus correcting the inordinately low ("significant") E-values for sequences with similarly biased amino acid composition. This improves the accuracy of the reported E-values and eliminates most false-positives. Composition-based statistics is currently used as the default for the NCBI BLAST. We will discuss the effect of this procedure on database search outcome in greater detail and using specific examples.

Sequence Assembly

Sequence assembly is a process whereby the order of multiple sequenced DNA fragments is determined. It signifies to aligning and merging fragments from a longer DNA sequence in order to reconstruct the original sequence. Advances in sequencing technologies and increased access to sequencing services have led to renewed interest in sequence and genome assembly. Concurrently, new applications for sequencing have emerged, including gene expression analysis, discovery of genomic variants and metagenomics, and each of these has different needs and challenges in terms of assembly.

Genome sequencing is a discipline that has undergone tremendous development in the past. With the introduction of the different new massively parallel sequencing technologies the field will go through further transformations as new challenges arise. Today 567 bacterial genomes with up to 10.5 million base pairs (Plesiocystis pacifica SIR-1) have been sequenced and submitted to NCBI. In addition several eukaryote genomes with approximately three billion base pairs have been sequenced and assembled, and many other sequencing projects are under way.

The experimental technique used in most de novo sequencing projects of higher organisms, DNA chain termination, was developed three decades ago and remains, except for much higher levels of automation, basically the same. Late in the 1970s the bacteriophages phiX174 and Lambda were among the first genomes to be completed together with the human mitochondrion.

In the following decade the shotgun sequencing strategy was introduced, and during the subsequent years it was extended by applying it to larger and larger DNA sequences cloned in plasmids (a few kilobases (kb)), cosmids (40 kb), artificial chromosomes cloned in bacteria (BAC – Bacterial Artificial Chromosome) and yeast (YAC – Yeast Artificial Chromosome), with inserts of 100 to 500 kb . The assembly of whole genome shotgun sequencing data was deemed to be futile until the successful WGS assembly of the 1.8Mb genome Haemophilus influenzae in 1994.

“High throughput” sequencing (HTS) of cDNA was initiated in 1991 by Adams, who also introduced the term “Expressed Sequence Tag” (EST) to refer to this new type of sequence information. Collections of ESTs have given a first good approximation of the diversity of all protein coding genes in a tissue. During the years ESTs have become an important tool with many applications, mostly in relation to gene analysis and gene discovery.

The amount of data generated by the different sequencing projects is overwhelming. For example, sequencing of the human genome produced 23 and 27 billion bases of raw shotgun sequences in the International Human Genome Sequencing Consortium and the Celera projects, respectively. However, the vast amount of fragments cannot readily be concatenated to a final sequence. Only by using computers it becomes possible to carry out the assembly of the pieces, but the outcome as well as the reliability of the result for a given type of data depends on the underlying strategy implemented in the computer program. Some strategies might be more suited for one type of data than others. Also, the computational resources of some methods might not scale well with the number of sequences in the data set. Though the experimental techniques have essentially driven the computational aspect of sequence assembly, the computational aspect is still of utmost importance since any meaningful assembly needs to is computer assisted.

One of the first assemblers introduced by Staden in 1980 was a computer program developed to store and manipulate DNA gel reading data obtained from the shotgun method of DNA sequencing. During the next decade several other programs were presented, among them SEQAID, CAP, PHRAP, and the TIGR assembler, which was used to assemble the genome of Haemophilus influenzae . In order to assemble larger and more complex eukaryotic genomes, new assemblers have been designed and implemented. Among them the Celera Assembler (now part of AMOS) and GigAssembler, both applied to human genome data sets; the JAZZ-assembler, which was applied to both the genome of Takifugu rubripes (the pufferfish) and Ciona intestinalis ; and the ARACHNE and Phusion assemblers, both applied to the mouse genome.

Several specific efforts have been undertaken in the context of EST assembly, and several tools are available. Among them are StackPack, TIGR TGICL, and geneDistiller. Some of the tools deal with splice variants or other problems such as chimerism (and includes alternative splice variants detection). Approaches to incorporate rather than remove repetitive sequences are discussed in.

Along with the increasing number of completed genomes, efforts are also made in developing computational methods for comparing genomes. These include TIGRs MUMmer, TWINSCAN,

GENEWISE, GENOMESCAN, BLAT, and AVID used for alignment and comparison of whole genomes, and FOR Repeats which is used to detect repeats on entire chromosomes and between genomes.

The massive effort to sequence the human genome produced a first draft version in 2001, and did, as a draft sequence, contain numerous gaps. It took another 3 years of sequencing and assembly before the finished version was presented (which still contains more than 300 gaps).

Sequencing Approaches

As mentioned the choice of assembly strategy depends on the sequencing method, and the choice of sequencing method may also depend on the organism that is being sequenced. Issues that can affect the final assembly (other than the obvious quality of sequence data) are the size of the inserts, whether the sequencing was uni- or bi-directional, the library construction, the cloning vector, the selection of clones to be sequenced, and the availability of additional information (consensus genome, ESTs, known verified genes, gene maps, etc).

Basic Sequencing Procedure

The basic procedure in sequencing has been to isolate genomic DNA or RNA (reverse transcribed into cDNA), and clone it into vectors (eg. plasmids, BACs) capable of stable propagation in suitable host cells such as Escherichia coli, for a schematic illustration of a sequencing vector. Several cloning systems with insert sizes varying from hundreds of base pairs to megabases have been developed. The ideal clone library for genomic sequencing has the following features:

- The clones are highly redundant, covering the entire genome many times (typically 6–10).
- The clone coverage is random and not biased towards or against specific regions of the genome.
- The clones are stable, not subject to recombination or reorganization during the propagation process.

It should be noted that one of the major improvements of the new massively parallel sequencing technologies is that they do not rely on vector cloning prior to sequencing, and the concerns listed here are therefore not directly applicable to those technologies.

After propagation, the clones are selected and the sequencing is performed. An essential feature in sequencing is the attachment of quality values to the raw sequences. The quality values indicate the likelihood of each base call being correct. In the assembly stage the quality values will help to distinguish true DNA polymorphisms from sequencing errors and match end sequences of low quality.

In genomic shotgun sequencing, which typically uses a single individual DNA source, sequences sharing less than 98% identity are usually assumed to come from different regions of a genome (including different repetitive elements). In contrast, EST data is usually derived from a variety of sources representing the spectrum of polymorphisms in the original samples. These will usually include a number of erroneous polymorphism which are caused by sequencing errors inherent in single pass sequencing, a relatively high rate of insertions and deletions, contamination by vector and linker sequences and the non-random distribution of sequence start sites in oligo(dT)-primed libraries. Therefore, the degree of identity in overlapping sequences from the same gene will often be lower in EST projects than in genomic sequencing projects. In addition, the patterns of overlapping sequences caused by alternative spliceforms are different from those observed in a genomic shotgun project.

The major tool to gather sequence information was the method introduced by Fred Sanger in the second half of the 70'ties. It uses dideoxynucleotide triphosphates (ddNTPs) as DNA chain terminators. The classical Sanger approach is carried out in four independent DNA polymerase reactions. Besides the DNA template and deoxynucleotides (dNTPs) a reaction mix contains ddATP, ddCTP, ddTTP or ddGTP. Each reaction results in DNA fragments of different length terminating with the respective ddNTP. Electrophoresis of the fluorescence- or radio labeled fragments allows the recovery of the template sequence. Later, the use of dye-terminators made it possible to perform sequencing in a single reaction rather than four – the basic principle however remained the same. While the classical Sanger approach requires separate synthesis and detection steps, High Throughput Sequencing (HTS) technologies employ sequencing-by-synthesis and sequencing-by-ligation approaches, allowing for simultaneous synthesis and detection.

Shotgun Sequencing

Two approaches for genome shotgun sequencing can be distinguished: whole-genome shotgun (WGS) sequencing and hierarchical shotgun sequencing.

Whole Genome Shotgun

Sequencing using the whole genome shotgun approach basically means that the genome is randomly broken into pieces and cloned into a sequencing vector. The inserts are subsequently processed to generate sequences of bases (referred to as reads). During the mid-1990s several groups recognized that sequence information from both ends of relatively long inserts dramatically improves the efficiency of sequence assembly. In contrast to single sequence reads from one end of the shotgun clones pairs of sequence reads from both ends have known spacing and orientation. Exact knowledge of the length of the insert is not required to utilize the advantages of end sequencing in assembly, but good estimates of clone length will aid the assembly immensely.

Hierarchical Shotgun Sequencing

The 'Hierarchical shotgun sequencing' (also referred to as 'map-based', 'BAC-based' or 'clone-by-clone') approach involves generating and organizing a set of large insert clones (typically 100–200 kb each) covering the genome (a "minimal tiling path"), followed by separate shotgun sequencing on each clone. For illustration see Figure. It is possible to establish a tiling path of overlapping BAC-clones using only BAC fingerprinting technologies. However, knowledge of unique genome markers (e.g. ESTs or sequence-tagged sites (STS)) and their location in the genome map is of great help for organizing the BAC clones in the correct order. In hierarchical shotgun sequencing the sequence information is local; therefore the risk of long-range and short-range misassemble is reduced.

Mixed Strategy Sequencing

A strategy that can be used on large complex genomes is the 'mixed strategy sequencing'. The technique utilizes both hierarchical and whole-genome shotgun. The method combines a light (x1) BAC clone coverage of the genome, with whole genome shotgun sequencing. The BAC clones act as a basic framework for WGS sequence assembly. The method was successfully applied to rat genome.

Reduced Representation Sequencing

A variant of WGS is "reduced representation sequencing" (RRS), where one selectively chooses subsets of the genome to avoid sequencing the (often much) larger regions that are not of interest. In, SNPs were discovered by mixing DNA from many individuals, preparing a library of appropriately sized restriction fragments, and randomly sequencing 5 clones. Here, the choice of the restriction fragments effectively selects only a small subset of the human genome. Several approaches to RRS have been employed for plant genomes: Methyl-filtration (MF) sequences uses the endogenous restriction-modification system of E. coli to eliminate methylated DNA inserts, the RescueMu (RM) approach focuses on the gene-rich regions which are rich in mutator transposons, and High-Cot filtration avoids repetitive and low copy sequences due to differences in the relative rates of DNA re-association. Most of the Maize and Sorghum genomes have been sequenced using MF.

Many of the applications of the new high throughput sequencing platforms are based on various RSS strategies This includes electrophoretic size separation to enrich for small RNA molecules reduced representation bisulphite sequencing for genome wide methylation analysis; flow sorting of derivative translocation chromosomes for breakpoint mapping; enrichment of DNA-fragments bound to specific proteins by chromatin immunoprecipitation of fixed, sheared DNA, for identification of transcription factor binding sites (CHiP-Seq) enrichment of specific parts of the genome by multiplex PCR-amplification or by hybridization to custom made arrays e.g. for SNP discovery and in situ exon capture.

EST Sequencing

Expressed Sequence Tags (ESTs) are sequences representing genes which can originate from specific tissues. In EST-sequencing a single automated sequencing from one or both ends of cDNA-inserts is performed. This single pass approach is the major reason EST-sequencing is cost effective. For additional information, see e.g. and references therein.

In most cases EST sequencing projects are aimed at establishing partial sequences of transcribed genes rather than full length cDNA sequences. However, this approach features some special challenges such as common sequence motifs, alternative transcripts and paralogous genes are challenges that potentially impact the assembly quality.

Massively Parallel Sequencing

Recently, a number of new sequencing technologies have emerged. The development was initiated by 454 sequencing and followed by Solexa sequencing and others. The common feature of all these technologies is that they are massively parallel, i.e. they generate a large number of different sequence reads in a single run. The generated small reads are usually aligned to a reference genome, and further analyzed.

The methods generally use one variant or another of fixing many sequence fragments on a substrate, cyclically adding different bases with some – technology-specific – luminal characteristics, and recording an image at each cycle. Image analysis is used to recover the all sequences at once. Sequencing of all immobilized fragments thus proceeds in parallel.

Compared to traditional sequencing a large amount of sequence data is generated at a drastically reduced cost per base. The most important disadvantage of high throughput sequencing is the significantly reduced read length, which limits their application in de novo sequencing of complex genomes (e.g. due to repeats), at least using simple shotgun strategies. However, these new platforms have many uses in genome resequencing, especially if it is possible to align the fragments to an existing good quality reference genome.

Due to the amount of raw sequence data, high throughput sequencing is valuable in areas such as SNP finding. In EST sequencing, HTS technologies might enable a researcher to make accurate digital expression profiles, even including low abundance transcripts, and help detecting alternative splicing (depending on the platform chosen).

One of the key technologies that gave rise to the era of HTS, pyro sequencing, was introduced in 1998. This sequencing-by-synthesis method is at the very heart of GS FLX systems by 454 Life Sciences. The detection is based on pyrophosphates (PPi) released during the polymerase reaction. Sulfurylase converts PPi to ATP which is subsequently consumed by luciferase to emit light in the visible spectrum. In GS FLX systems, a library of DNA templates is immobilized on DNA capture beads, amplified using emulsion PCR (emPCR) and loaded onto proprietary titer plates with several hundreds of thousands reaction wells. During a run, the four nucleotides are flowed sequentially over the 6 plates. The luciferase reaction triggered by nucleotides complementary to DNA templates is recorded by a CCD camera. A washing step is necessary to allow the next detection step. The GS FLX currently allows read lengths of several hundred bases. According to the manufacturer a single instrument run with two high-density plates generates information for about 20 million base pairs.

A competing technology, Solexa, now sold by Illumina, uses optically transparent surfaces to immobilize fragmented and adapter-tagged DNA. Each attached fragment is subsequently amplified ~1000 fold by repeated steps of bridge amplification. The resulting clonal clusters are then sequenced using reversible terminators with removable fluorescent dyes. With approximately 30-40 bp Solexa reads are significantly shorter compared to GS FLX. However, close to 50 million clones per flow cell can be sequenced in parallel, resulting in presently >1.5 Gb of sequenced DNA in a single sequencing run. This amount can be doubled by sequencing the other end of each fragment (paired-end). Improvement in chemistry may further increase the read lengths and hence push the total amount of sequenced DNA well beyond the size of a human diploid genome.

A third synthesis-based technology, tSMS (true Single Molecule Sequencing), is currently distributed by Helicos. No DNA amplification is required for this approach. Instead, fragmented single stranded DNA molecules are directly immobilized on a solid surface. Similar to Solexa, tSMS works with nucleotides that carry a removable, laser light-detectable fluorescent. At the moment the system is able to sequence reads with lengths up to 55 bases at a speed of 25 to 90 million usable bases per hour.

SOLiD, a system now sold by Applied Biosystems, is a technology that uses a sequencing-by-ligation approach. An adapter-tagged library of short DNA fragments is amplified with emPCR, immobilized on capture beads and then deposited onto high-density glass arrays. The SOLiD sequencing-by-ligation protocol uses four by four sets of 8-mer probes. In each set only two bases, fluorescently labeled, are specific. The interrogation of sequences is done in four phases. If a probe has specifically bound to the free template in the first phase, say at position 1 and 2, it is enzymatically ligated to the current 5′ end at position 0. After the detection step 3 nucleotides of the probe along with the fluorescence label are cleaved. The next ligation step interrogates 6 and 7 and so forth. After the first phase, the ligated sequence is removed, and another set of bases are called. So in the second phase bases 2 and 3 are read, in the third 3 and 4 and so forth. The advantage of the SOLiD system is that the double base reading leads to an increased accuracy. Currently SOLiD produces read lengths of about 30-40 bp and a total of 9 Gb per single run, with read length expected to become longer in the future.

In the future, other technologies may become available, such as the use of solid state nanopores for sequencing of single DNA molecules. We refer to for an overview, in which several interesting ideas how this approach could be implemented in practice are presented.

Mapping of Short High-throughput Sequencer Reads

Compared to de novo assembly, the mapping of resequenced reads to a template genome is a computationally easier problem. Still, efficient mapping tools are crucial, and several tools for mapping of short reads are available. Most of the tools, i.e. MAQ, SOAP, SHRiMP or Eland (proprietary), use seeding techniques that gain their speed from precomputed hash look-up tables. Typically, seeds of fixed length allow for not more than one or two mismatches. In addition, the capability to detect insertions and deletions, as they frequently occur in 454 sequences is very limited, and most programs can only detect indels in subsequent alignment runs. For short sequences it would be helpful, but computationally more expensive, to incorporate indels right from the start. Current mapping tools have different additional features. The program MAQ, e.g., additionally supports paired-end read matching — helpful to deal with paired-end reads produced e.g. by the GS FLX and other high throughput platforms.

Computational Assembly

Computational assembly is the only way to efficiently assemble sequenced fragments of DNA. However, a sufficient amount of high quality sequences are required. The assembly programs should be able to handle large data sets effectively 7 and avoid misassemblies in the presence of large repetitive or duplicated regions and redundant sequences. To accomplish this, effective algorithms to handle large input data sets with the use of minimal computer time and memory are needed.

One of the primary difficulties in computational genome assembly is to develop an algorithmic approach capable of detecting stretches of repetitive DNA without causing misassemblies. Repetitive sequences complicate assembly as different pieces of sequence can share the same repeat sequence originating from different genomic locations. Since the pieces are put together by searching for matching overlapping nucleotides, repeats can be put together erroneously. Typically, for shotgun data, repetitive sequences are revealed by clusters containing more overlapping reads than would be expected by chance.

In EST datasets the main difficulty is to develop an algorithmic approach that, in addition to efficient assembly, can handle highly expressed genes, paralogous genes, alternative spliceforms and chimerism in the dataset.

Although pyrosequencing with a whole-genome shotgun approach has been successfully applied to bacterial genomes, the construction of high-quality assemblies with high-throughput sequencing data is still a non-trivial problem even for short genomes. At present, no approach has been proposed to directly assemble large animal or plant genomes directly from short sequences obtained using HTS. The Short Read Assembly Protocol (SHRAP), however, comprises a protocol for high-throughput short read sequencing that differs in two respects from classical hierarchical sequencing approaches. This protocol however, expects read lengths much longer (200 nucleotides) than those produced by SOLiD or Solexa. The assembly methodology is based on the Euler engine introduced in 2004. The Euler-SR assembler, specifically designed to assemble short reads, uses an updated version of the Euler engine to reduce memory requirements. The results for real Solexa reads, however, were less convincing due to the poorly understood error model and highly variable error rates across different machines and run times.

Basic Principles of Assembly

For the majority of traditional assembly programs the basic scheme is the same, namely the overlap-layout-consensus approach. Essentially it consists of the following steps:

- Sequence and quality data are read and the reads are cleaned.
- Overlaps are detected between reads. False overlaps, duplicate reads, chimeric reads and reads with self-matches (including repetitive sequences) are also identified and left out for further treatment.
- The reads are grouped to form a contig layout of the finished sequence.
- A multiple sequence alignment of the reads is performed, and a consensus sequence is constructed for each contig layout (often along with a computed quality value for each base).
- Possible sites of misassembly are identified by combining manual inspection with quality value validation.
- Prior to the assembly, the electropherogram (for Sanger sequencing, images for massively parallel sequencers) for a given sequence is interpreted as a sequence of bases (a read) with associated quality values, these values reflect the log-odds score of the bases being correct. The basecaller PHRED is often used, however alternatives exist, e.g. The CATS base caller.

The reads can then be screened for any contaminant DNA such as Escherichia coli, cloning or sequencing vector. Low quality regions can be identified and removed. Base quality values can be used in computation of significant overlaps and in construction of the multiple alignments.

For high-throughput sequencing data, the basic proposition for SHRAP is to sample clones from the genome at high coverage, while sequencing reads from these clones at low coverage. SHRAP starts off with assembling the reads greedily to small local assemblies and subsequently to contigs on each clone. It proceeds by ordering the clones in a "clone graph", and constructing "clone contigs", which are then assembled independently. Computer simulations of the procedure show that the approach can reach a quality comparable to the current assemblies of single human chromosomes and fruit fly genomes using reads of 200nt with an error rate of not more than 1%. These are constraints that are too strict for short (Solexa or SOLiD) reads (≈ 40bp) and because of higher error rates challenging for real 454 reads. Furthermore, for mammalian genomes the use of a hierarchical sequencing strategy might be somewhat cumbersome.

However, the use of templates might bail Solexa and SOLiD users out: In a recent study, de novo assemblies of chloroplast genomes (≈ 120 kb) were improved by aligning preassembled contigs to reference genome. After de-Bruijn graph assembly of reads, small contigs were aligned to closely related chloroplast genomes. Between 67% and 98% of the contigs could be aligned to such templates. If alignment failed, sequences were scanned for similarity using BLAST. In the end, however, their assemblies were estimated to be 88–94% complete. Yet, the assembly of mammalian genomes or genomes without good reference sequences seems to be a considerably more difficult task. The successful de novo assembly of Chloroplasts genomes with 454 reads has been shown earlier.

454, SOLiD, Solexa technologies allow convenient generation of mate-pair/paired-end sequences, i.e. the ability to sequence both ends of each DNA fragment. However, in an assembly using a hybrid dataset of real 454 reads and simulated mate-pair data, about 96% of the mate-pairs did not contribute additional information and hence did not improve the assembly. Likewise, in a hybrid dataset of 454 reads and Sanger reads the vast majority of long sequences did not improve the assembly substantially, measured by N50 contig size. Despite those simulation results, the latter method has already been shown to work quite well in practice, and one area where mate-pair/paired-end sequencing should improve the analysis dramatically is for the detection of breakpoints related to structural rearrangements, e.g. deletions, duplications, inversions and translocations.

General Assembler Differences

When different assemblers try to piece the DNA puzzle together they essentially work from the same input, but the assemblers differ in the way they utilize the sequence information, and in the way this is combined with additional information. In general the differences fall in the following categories.

Overlaps: A lot of different methods are used to find potential overlaps between sequences. Some are based on BLAST (e.g. gene Distiller), while other assemblers use various other methods to find similarities between reads:

- Additional information: Depending on how the sequence reads are produced some additional information might be available. This information might consist of read pair information, BAC clone information, base quality information, etc. Some assemblers use this data to impose additional structure on the assembly of the sequences (e.g. Gig Assembler).

- Short read assembly: De novo assembly of the micro reads generated from next generation sequencing platforms is still challenging. While assemblers have been developed and applied to assemble bacterial genomes successfully, on larger genomes the assembly is performed by mapping the micro reads to reference genomes. The major next generation sequencing platforms all have built-in software to handle this task, e.g. GS Reference mapper, Gerald for Solexa. In SOLiD systems the mapping tool "map reads" converts reference sequences into color space and perform the mapping in color space.
- A somewhat related issue is how the sequences are cleaned of contaminant sequences (ie. vector sequences, repeat sequences, etc.). While this can essentially be considered separately and independently from the assembly itself, some assemblers incorporate cleaning in the way they process the reads (e.g.).

Overlap Identification and Alignment

In a whole-genome context, trillions of overlaps between reads are examined. The majority of assemblers use alignment algorithms which are general modifications of methods first introduced by Needleman and Wunsch in 1970, Smith and Waterman in 1981 and Gotoh in 1982.

Initial overlap detection is often performed by finding exact identical subsequences (often called words, k-words or k-mers) between reads, prior to making the actual alignments. These identical subsequences are used to find pairs of potentially overlapping sequences, which can then be aligned to each other in order to check if they represent a true overlap. The size of the subsequences varies from method to method, and is dynamic in some assemblers. Furthermore, the identical subsequences are grouped and used in different ways depending on the assembler.

For almost all assemblers, a modified Smith-Waterman algorithm is used to align candidate overlapping reads. The basic idea in the alignment algorithms is to use dynamic programming to construct a matrix containing scores of all subsequences, which is then analyzed to find the "optimal" alignment. Dynamic programming simply means that the alignment is calculated as extensions to already aligned subsequences. The assemblies programs differ in their exact implementation of this algorithm, as (nearly) all of them use a heuristic approach to decrease the computational load, thereby increasing speed (e.g.). In the assembly of ESTs a clustering step is used to group the input sequences sharing significant regions of near identity together.

Multiple Alignments and the Consensus Sequence

While the alignment of two sequences is usually straightforward, aligning more than two is not so simple. The standard Smith-Waterman algorithm can easily be extended to the task of aligning many sequences by constructing a "multidimensional matrix". However, the number of calculations rises exponentially with the number of sequences. This sets severe practical limits of the number of sequences that are viable to align, and therefore finding the true sequence from a number of overlapping reads becomes difficult.

Precisely how the different assemblers generate a multiple alignment and consensus sequence is only vaguely but a common approach is to use a heuristic greedy algorithm. The greedy algorithm typically performs pairwise alignment between overlapping reads, from which a multiple alignment is build up iteratively, i.e. adding one sequence at a time, but with this approach there is no guarantee that such a multiple alignment is correct.

After the multiple alignments have been constructed the consensus sequence is found. This would typically be the sequence generated by taking the most common base at each position in the alignment, however other methods exist. For instance geneDistiller, where ungapped alignments of reads is performed (thus simplifying the multiple alignment). The consensus sequence is constructed by splitting the multiple alignments in 12-mer words and analyzing the relative frequencies of these, where the presence of alternative transcripts is detected through the frequencies of the 12-mers (and displayed as stretches of 'alternate consensus').

The assumption is that the final consensus sequence corresponds to the original genomic sequence where the sequenced fragment originates.

Eulerian Fragment Assembly

In assemblers aimed at short read assembly (e.g. SOLiD reads) an approach based on mathematical graph theory is often used, namely the Eulerian fragment assembly method. The Eulerian fragment assembly avoids the costly computation of pairwise alignments between reads. The De Bruijn graph of a genome has as its vertices all distinct $k - 1$ tuples that occur within the sequence (where k is the word length that is used). A directed edge is inserted between s and t if there is a k tuple $(u_1,u_2,\ldots,u_k-1,uk)$ in the genome such that $s = (u1,u2,\ldots,uk-1)$and $(t = u_2,\ldots,uk-1,uk)$, i.e., if s and t appear shifted by single nucleotide. In practice one uses the k-tuples appearing in the collection of the sequence reads and a value of k between 6 and 9 or 10. In the error-free case, the genomic sequence can be read off directly as an Eulerian path through the De Bruijn graph (with repeats forming "tangles"). In real, error-prone data underrepresented k-tuples, i.e. K-tuples that appear less frequently than expected from the coverage rate, indicate sequencing errors and can be omitted.

Data Reliability

Preprocessing and Cleaning

A critical aspect of any large-scale sequencing effort is the production of high quality data. To obtain this preprocessing is applied to the reads. For Sanger sequencing this includes base-calling, filtering of low quality reads, short length reads (typically less than 100 bp), identification of sequence features such as linker restriction sites, cloning vectors, polyadenylation tails, library tags, polyadenylation signals and other contaminants like bacterial sequences.

There are different computational programs available to detect these contaminations. Most of the existing programs used for processing solely focus on a single step. While PHRED deals with base-calling, cross match aims to identify and mask vector sequences in reads. Preprocessing can also be done using other programs such as LUCY, a sequence trimming script like SeqClean, or ESTprep.

In the Solexa system, the module for sequence alignments, Gerald, applies some filters to remove low quality base calling before the real mapping starts. As it is based on optical detection of ultra-high dense sequence clusters on surface, chastity and purity of optical signals are crucial for accessing the quality. Distance between clusters is also taken into consideration. Thresholds for these features can be customized in the program. Other next-generation sequencing systems employ different measures according to their methods.

Repeats

In mammalian genomes the repetitive content can be as high as 50%. The repeated fraction includes interspersed repeats derived from transposable elements, and long genomic regions that have been duplicated in tandem, palindromic or dispersed fashion, e.g. ribosomal RNA genes, centromeres, heterochromatin and retrotransposons. Such features complicate the assembly into a correct finished genome sequence and have a great influence on the design of assemblers. Computationally repeats are typically handled as follows:

- Comparing: By comparing reads to known repeated regions in other genomes, potential repetitive sequences can be separated (and typically discarded) from the assembly.
- Masking: Regions which have a high depth, that is regions where many reads share the same sequence, are marked as repeats. Usually such regions are discarded by the assembler, and are not incorporated in the assembly, e.g. by the method presented in . A standard program for masking repeats is RepeatMasker. It searches through curated repeat databases (eg. Repbase) using the alignment program cross_match to identify and mask repeats. The speed of cross_match can be increased by using the software wrapper MaskerAid.

Expressed Sequence Tags

Due to the way that the EST sequences are generated, there are several concerns which can severely disrupt attempts to analyze the data:

Over-clustering: This happens when ESTs from different genes are clustered together, and therefore associated with the same genetic sequence. This often arise as a result of the cloning procedure, which falsely place two originally separate sequences in the same read, ie. chimerism. However, paralogous genes can also be clustered together due to high sequence similarity. Using the traditional (TGICL, d2_cluster) single transitive single linkage clustering methods can cause all EST from both genes to be assigned to the same cluster. More stringent clustering methods such as the double linkage of geneDistiller can reduce the amount of falsely clustered reads, and create more consistent assemblies and consensus sequences.

Highly expressed genes: In non-normalized cDNA libraries the fraction of the genes that is highly expressed, will be represented in a high number and lead to large and deep clusters that may accidentally contain EST from more than one gene. There are several ways to handle highly expressed genes depending on the purpose of the investigation:

- Removal of known housekeeping genes: If the sequence of some housekeeping genes of the organism is known, removing ESTs that originate from these genes can alleviate the problems.
- Adding annotated gene sequences: If a genetic sequence of an annotated gene is known, it can be used as a template for the ESTs.

- Seeded clustering: Known full-length transcripts can be used for 'seeded clustering', which helps to create smaller, better partitioned clusters and avoid chimeric assemblies.

These procedures can alleviate some of the problems; however some clusters of highly expressed genes can still contain several thousand EST sequences. Producing a consensus sequence from such a large cluster can be tricky as most assembler is not able to handle such deep clusters. Several methods have been created to deal with this problem, such as the "containment clustering" of TGICL, or the alignment/consensus strategy of geneDistiller.

Other minor concerns in EST assembly are overlapping genes e.g. they can be on opposite strand and share a UTR-tail or have common motifs. This can cause the assembly program to assign ESTs from two different genes to the same cluster, and will complicate analysis of the cluster.

Reliability of high-throughput assemblies and sequence data. In a survey of assemblies for Streptcoccus suis from 454, Solexa and capillary data, 454 sequencing of a library with 5-fold coverage produced 5336 contigs while the Sanger method, two-fold coverage, resulted in only 1011 contigs. The length of the largest contig was 5336 for 454 and 12257 for the capillary sequencing method. Moreover, using Solexa, a ten-fold coverage was necessary to produce 8370 contigs with a maximum length of only 1687. The best results were seen for hybrid assemblies comprising data from at least two different sequencing technologies. The differences are likely to be caused by very different error patterns. In the case of the Solexa technology, error rates are highly position-dependent, variable across different machines and even across different runs. In a recent investigation on the quality of Solexa reads, significantly more reads were found in GC-rich genomic intervals. Despite the manufactures specifications for the read quality, error rates varied from 0.3% to 3.8%. Compared to 454 sequences, only few insertions and deletions were found. In the future a new base calling software, e.g. Alta-Cyclic, might be able to improve the quality of Solexa sequences. Additionally, it has been shown that under idealized conditions it is theoretically possible to assemble bacterial genomes (with 80x coverage of 30 nt reads).

Assembly of Contigs - Scaffolding

While the assembly of the individual reads into contigs gives some (local) information, the contigs still need to be set into the context of the whole genome. This is carried in the last phase of an assembly process: scaffolding, which is the process 12 where different (genomic) contigs are organized into even larger frameworks (scaffolds or super-contigs). The contigs are ordered and oriented in a consistent way, so that the scaffold build is a true representation of chromosomes, though there may still be gaps between contigs, which are dealt with in new rounds of sequencing.

In the scaffolding stage of an assembly, all the information usually comes from other sources than the reads themselves. This information includes read-pair information, STS (Sequence Tag Sites), and other sources.

Finishing

When an assembly has been completed, specific parts of the assembly usually need to be reexamined, perhaps due to low quality of the data, low (or no) coverage of the sequence, sites under suspicion of misassembly, etc. The reexamination is usually dealt with in an elaborate process where manual inspection is used to analyze the ambiguous section(s) and new ways are devised to clarify the particular ambiguities.

Analysis of the assembled contigs can be performed with a number of tools. One is Consed, which allows navigation of the assembled contigs and reads, problematic regions can be searched for with different criteria, and regions can be tagged for further inspection. Others are Autofinish, BACcardi, and GAP4, all of which has different strengths and purposes.

Genome Resequencing

Recent developments in high-throughput sequencing technologies have ignited the scientific community's imagination. Terms such as the "personal genome"or "1000$ genome" are now popular in the media. The growing number of publicly available reference genomes allows genome resequencing on a larger scale, as sequencing costs decrease and throughput increases. However, currently even HTS only allows deep resequencing of a small number of large individual genomes or of specific parts of the genome. It has been remarked that the full power of high-throughput sequencers might not be unleashed since no suitable methods to select for specific genomic subsets are available and methods for targeted amplification are more likely to be effective. However, recent methods using hybrid techniques such as microarray-based genomic selection (MGS) and multiplex exon capture to narrow down the number of sequences or to focus on specific genomic locations may overcome this shortcoming. In this project a set of 106.5 million reads, representing 24.5 billion bases and a depth of 7.4-fold, was generated. The mapped reads in combination with 454 quality values (Q-values) were used to gather a set of 3.32 million SNPs. Several filters had to be applied to increase specificity.

A read was only included if the BLAT alignment:

- Was spanning at least 90% of the read length.
- Did not have alternate hits.
- Had less than five mismatches.
- Had less than five indels.

Subsequently, the remaining 93 million reads were again realigned with PHRAP's cross_match tool. Three additional filter steps using the quality score, (see supplementary material for), the ratio of the variant to total coverage (> 0.2) and the vicinity to homopolymer runs (< 5bp) in order to avoid false positive indels ended this complicated procedure. Additionally, approximately 2.6 million reads of novel sequence and reads with low quality alignments were assembled in 170 000 contigs spanning 48Mb. After a filter step 110000 contigs spanning 29Mb remained. With costs of about 1 million US$, however, the "1000$ genome"genome still seems to be a distant prospect.

Next-generation HTS has also been applied for the mapping of translocation breakpoints. HTS not only reduces the labor and time cost of traditional methods in detecting translocation breakpoints, eg. in situ hybridization with fluorescent dyelabeled bacterial artificial chromosome clones (BAC-FISH), but also greatly improve the resolution so that the disrupted gene can be identified by PCR cloning. Thus, mapping and sequencing breakpoints region with Solexa platform has been used to identify novel candidate genes for mental retardation. Probability calculations as well as simulations suggest that current paired-end sequencing technology already provides a high probability of breakpoint detection and good resolution in localizing structural chromosomal the rearrangements.

Assembly Methods

Different assemblers use different information in the assembly process. Some only use sequences in fasta format and the corresponding quality values, while others can assemble without quality values. Additional information on known sequences (e.g. genes), clones, clone sizes and the orientation of the reads (forward-reverse) might be helpful in the assembly process.

Assemblers

In the following a large selection of different assemblers that have been created over time are presented. One of the (relatively) early assemblers is PHRAP, which is still in use, both in itself (for small DNA sequence sets), and as a subcomponent of WGS assemblers, eg. RePS, Phusion, JAZZ, and ATLAS . Other WGS assemblers that also use some variety of the standard overlap-layout-consensus approach are, the Celera assembler, CAP3, RAMEN, PCAP, the TIGR assembler, STROLL, and ARACHNE2. Some new approaches to assembly have been attempted, among them mira and TRAP, which try novel ways to deal with repetitive sequences by checking the trace and quality files. An emerging approach is to use more explicit graph based programs, such as Euler, Partial ordered alignment (POA), Velvet, Splicing graphs, ASmodeler, and xtract, where the last three are used specifically for ESTs. Other programs that analyze ESTs are TGICL, Stack Pack, PaCE, Hidden Markov Model (HMM) Sampling, and geneDistiller . Finally, some programs are used in the scaffolding stage, where contigs are processed and put in order, eg. Gig Assembler and Bambus (part of the AMOS package).

Assembler Comparisons

Comparing the different assemblers is not a trivial task due to several factors. Not to mention the problems of constructing appropriate benchmark data. First the different assemblers use a variety of input data, and so comparing an assembler which uses a lot of the additional information to one which only uses a fragment of the information is inappropriate. Another aspect is evaluating the success criteria; the goal is to create a single error-free contig of each chromosome, which means that fewer gaps, longer contigs, and fewer errors are desired. However, different assemblers might do better in one area and worse in another, so weighing the performance of one assembler against another can be difficult. Still there have been a few attempts to compare assemblers.

In, PHRAP, TIGR Assembler, and STROLL were compared on sequence data from the bacterium Borrelia burgdorferi. Phusion and ARACHNE were both applied to the assembly of the Mouse genome. PHRAP has been compared to CAP3 in (on four BAC datasets) and (on EST data) where the TIGR Assembler was also included.

Common to these studies is that the individual performance of the assemblers depends on the data they are presented with. PHRAP is generally aggressive in joining reads and creates large contigs, though sometimes at the expense of introducing errors. This assembler would be a fairly good choice if the dataset consisted only of reads with assigned quality values. However if additional information, such as forward-reverse constraints, is available other programs (e.g. CAP3, STROLL) would perform better. Another observation is that the performance of PHRAP degrades when it is applied to some large data sets. Additionally an updated assembler based on the Euler package, Euler-SR, is available. Euler-SR 14 which uses a revised version of the Euler package is less space intensive and optimized for short Reads. Alternatives are assemblers such as Arachne or Velvet.

Applying Assemblies for other Analyzes

There are different possibilities for further processing of the data and thereby for finding interesting and important features for future investigation, for example searching for SNPs (Single Nucleotide Polymorphisms) and alternative splice forms, or comparing genomes with each other.

SNP detection: ESTs are the most often used data source for SNP detection, but SNPs can be found from shotgun data as well. SNPs in transcribed sequences can either be synonymous (no amino acid change), or non-synonymous (encoding a different amino acid). A variety of different computer programs are designed for SNPs analysis. Some find and predict whether a given site is polymorphic, e.g. Polybayes, Polyphred and novoSNP. Others try to predict whether a given SNP is potentially harmful or neutral, e.g. Polyphen and SIFT.

Massively parallel sequencing the new massively parallel sequencing technologies will provide a wealth of new information. They have already been applied for the sequencing of an individual's genome, and detection of genomic rearrangements, and in the future new ways of utilizing their enormous capacity will likely appear, both with respect to the number of clones that are analyzed and the total amount of sequenced DNA.

Detection of alternative splicing: In eukaryotes, the removal of introns by splicing is a crucial step in gene expression. For some genes, splicing results in only one single type of mRNA, but studies have revealed that up to 60% of the human genes result in two or more mRNA isoforms due to alternative splicing. One approach to investigate alternative splicing is through assemblies of ESTs. However, assemblies of ESTs usually has multiple solutions in the presence of alternative splicing, which might end in truncated, misassembled or missing transcripts. Having a completed genome as a reference can help because it allows comparison of the EST to the corresponding genomic sequence. Some programs have been created which explicitly try to address the problem of assembling alternative splice variants from ESTs, among them are Splicing Graph and geneDistiller.

Genome Comparison: Furthermore, as different sequencing project complete their respective genomes and the data become available, it becomes possible to compare differences and similarities between different species on a sequence basis. This can generate a wealth of new information, and give new insights into the evolution and biology of living organisms.

Sequence Alignment

Sequence Alignment is a process of aligning two sequences to achieve maximum levels of identity between them. This help to derive functional, structural and evolutionary relationships between them.

Aligning sequences assigns functions to the unknown proteins, determines the evolutionary relatedness of organisms and helps in making prediction about the 3D structures.

Homology is attributed to similarity due to a descent from a common ancestor, i.e. if two sequences from different organism are similar there is a possibility that both sequences are termed as Homologous. Thereby, predicting structure and function for the sequences.

Types of Sequence Alignment

- Based on sequence Length: According to the length of sequence being compared it is of following two types:
    - Global sequence Alignment: In this alignment sequences are aligned along their entire length to include as many matching characters possible.

```
Seq 1 -   TAGC-GC-GT
Seq 2 -   TA-CA-CAGT
```

    - Local sequence Alignment: In this alignment sequences are aligned to find a region of higher density or strong similarity.

```
Seq 1 -   CGATAACGTAT
Seq 2 -   --ATAAAC---
```

- Based on Number of sequence: According to number of sequence being compared it is of following two types:
    - Pairwise Sequence Alignment: This involves aligning two sequences and to get the best region of similarity.

```
Seq 1 - 1 KTSSGNGAEDS    11
          |||||||||||
Seq 2 - 1 KTSSGNGAEDS    11
```

Various methods used for pairwise alignment of nucleotide and protein sequences are:

- Dot Plot: It is graphical method for two sequences to identify the region of maximum similarity and dissimilarity, depicted by presence and absence of DOTS.
- Dynamic Programming: This method breaks a problem into small sub-problems and uses the solution of the sub-problems to compute the solution of the larger one. Some algorithms like Needleman-Wansch and Smith-Waterman are used here.
- Heuristic Method: When a single sequence is to be compared against the whole database heuristic methods like BLAST and FASTA are used.

The following are certain parameters used for producing optimum alignment:

- Max target sequences: It displays the result with total number of aligned sequences on a page.
- Expected Threshold: It is a statistical indicator which calculates the probability that the resulting alignment are caused by random chance. The lower the E value, the more significant is the score The default value is kept 10 as 10 matches are expected to be found random by chance.

- Query match: It gives the maximum match in a query range. This is useful for comparing many stronger matches of the query results from weaker matches of the results.
- Word size: This algorithm works by using word matches between the query and the database sequences. It searches for exact word match, initiates the extension leading to the full alignment. Word size 3 is required for standard protein align. Word size 2 is required for short and nearly exact matches.
- Scoring schemes: Different scoring schemes algorithms are devised to obtain an optimum alignment. Use of any substitution matrix helps in aligning possible pair of residues and also generates scores. To check the quality of pairwise sequence alignment, different PAM and BLOSUM matrices are used.
- BLOSUM (Block amino acid substitution matrix): This has been developed using conserved regions called BLOCKS. Of distantly related protein sequences available from the block database. Out of all BLOSUM 62 matrix is best used for detecting most protein similarities. BLOSUM 45 may be used for longer and weaker alignments.
- PAM (Point Accepted Mutation): This is developed by calculating the substitution of amino acid during evolution which are naturally accepted. PAM 30 is used for sequences less than 35 in length whereas PAM 70 is used for sequences ranging from 35 to 50.
- Gap costs: A gap is a space which is introduced into an alignment to compensate for insertions and deletions in one sequence relative to another. Too many gaps should be avoided in the alignment and hence a gap penalty or gap score is assign. The introduction of gap causes the deduction of gap score from the alignment score.

Extension of the gap to encompass additional nucleotides or amino acid is also penalized in the scoring of an alignment. Increase in gap costs parameter results in the decreased number of gaps in the alignment. The penalty for the creation of a gap should be large enough so that gaps are introduced only where needed, and the penalty for extending a gap should take into account the likelihood that insertions and deletions occur over several residues at a time. Some used values here are existence 10, 11 and extension 1.

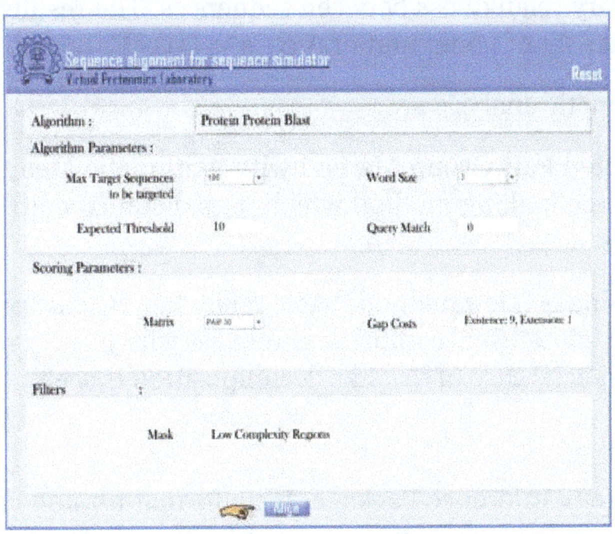

- Multiple sequence Alignment - This involves the alignment of more than two (protein, DNA) sequences and assess the sequence conservation of proteins domains and protein structures. It is an extrapolation of pairwise sequence alignment which reflects alignment of similar sequences and provides a better alignment score.

Various analysis like Homology modeling for prediction of protein structure, Phylogenetic analysis, motif detection etc are based on the results of multiple sequence alignment.

There are many softwares like Clustal, t-coffee, Phylip, MSA, MUSCLE used for obtaining multiple sequence alignment. Example –

```
Seq 1 - PQGGGGWGQ

Seq 2 - PHGGGWGQ

Seq 3 - PHGGGWGQ

Seq 4 - PHGGGWGQ

Seq 5 - PHGGGWGQ
```

Following parameters should be considered to align multiple sequences:

- Protein weight matrix - Matrix is used to increase the alignment score. Eg: PAM and BLOSUM.
- Gap Open – The penalty to open a gap. The presence of a gap is frequently given more significance than the length of the gap. By default, the gap opening penalty is 10.
- Gap extension – The penalty to extend a gap. Extension of the gap also involves additional amino acid penalized in the scoring of an alignment. By default, gap extension penalty is 0.20.

Application of MSA Results

Phylogenetic Analysis – It is one of the major areas where multiple sequence analysis results are used to find the evolutionary relatedness between sequences. The results are displayed in form of Phylogenetic tree which has set of nodes and branches to link the nodes.

Methods used for Phylogenetic analysis are:

- UPGMA (Unweighted Pair Group Method with Arithmetic Mean) – it is a simple, hierarchical clustering, tree making method which uses distance matrix to find the relatedness between sequences.
- NJ (Neighbor Joining) - It is a method that is related to the clustering method. The method is especially suited for datasets comprising lineages with largely varying rates of evolution. In this method, a special case of the star decomposition is seen.

File format view:

- PHYLIP - PHYLogeny Inference Package. It's a format for Joe Felsenstein's phylogenetic applications, having 8 letter maximum lengths for the sequence ID.

- Claudogram - In a cladgram, the external taxa line up neatly in a row Their branch lengths are not proportional to the number of evolutionary changes and thus no Phylogenetic time analysis can be done only the relative ordering of the taxa can be analyzed.

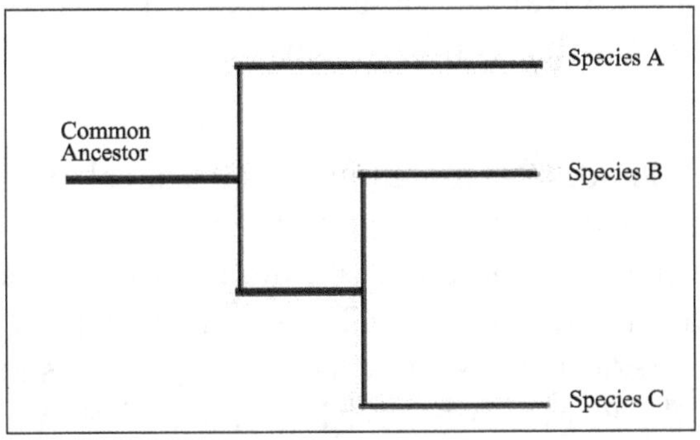

- Phylogram – In a phylogram, the branch lengths represent the amount of evolutionary divergence. Such trees are said to be scaled.

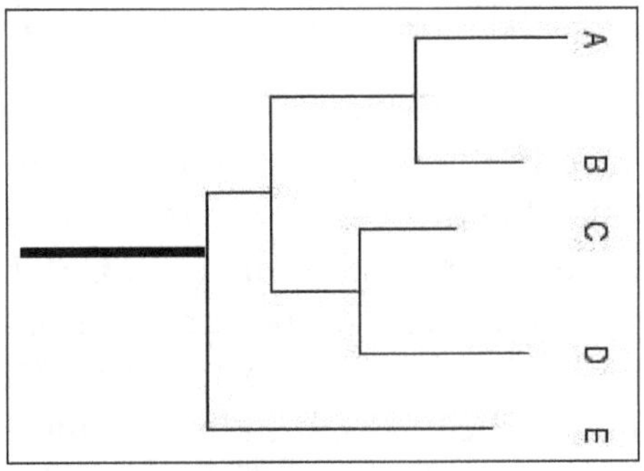

- Pearson/ FASTA – Text based format to represent amino acids in single letter code. It also has sequence names followed by comments.
- Jalview – Java alignment editor. It is a visualization tool for alignment algorithms and other database search results.

Pairwise Sequence Alignment

Pairwise alignment is one of the most fundamental tools of bioinformatics and underpins a variety of other, more sophisticated methods of annotation. Pairwise alignment in its most rigorous form uses a method called 'dynamic programming', which is highly accurate, but also incredibly costly to compute.

In order to align anything other than an exact alphabetic match, the algorithm has to know what it is looking for and how it can evaluate the worth of what it finds. To this end, 'comparison matrices' have been created which define a score for every possible match possibility—an effective tally of how well the computational alignment is doing. The software will search for the highest score available. The final score is relevant only with its resulting alignment and cannot be used outside this context.

In the case of DNA, comparison values are generated using a simple identity matrix of the type that allows one (positive) score for a correct match within the alignment, and a different (zero or negative) score for a mismatch. A fuller comparison matrix allowing ambiguities, alters these basic values as potential transitions and transversions are taken into account—but essentially there is very little mathematical difference that can be achieved between one alignment and another similar one.

Protein matrices, on the other hand, offer a greater breadth of calculation as not only are there five times as many common amino acid residues as there are DNA bases; they incorporate a significant amount of evolutionary information. The most common matrices here are the position accepted mutation (PAM) and BLOSUMomparison tables. The first PAM matrix was created in the late 1970s and relied on noting accepted residue substitutions within protein sequences to produce the PAM 1 table. Subsequent tables in the PAM family have been created by multiplication models based on that first matrix. The greater the multiplication, the higher the number in the PAM series and the greater the number of accepted mutations which have been involved in the proteins used to create the tables, and thus the greater the evolutionary divergence of those proteins. One of the more common matrices—the PAM 250 matrix—represents a subset of proteins of approximately 80% diversity.

The BLOSUM matrices were created in the early 90s and relied on the presence of residues within the blocks of conserved regions of related proteins to create the matrix. These blocks can be accessed in the BLOCKS database. There is also a family of BLOSUM matrices which are differentiated with numbers. These numbers, however, represent the minimum percentage identity of the BLOCKS used to create the matrix. The most common matrix of this set is BLOSUM 62, a default setting for many protein alignment applications. It indicates that BLOCKS of at least 62% identity were used in the creation of this matrix. In the case of the BLOSUM matrices, the higher the number connected to the BLOSUM matrix, the smaller the evolutionary divergence between sequences.

Once the comparison matrix has been established, the computer can make its own matrix based on the two sequences to be aligned—inserting a 'score' for each potential base or residue alignment. However, to allow the computer to score each comparison and select the one match—or run of matches—that gives the highest score would not necessarily yield the best alignment as it ignores biological insertions and deletions. In order to accommodate an insertion, gaps must be tolerated and as with the bases or residues, they must contribute a numerical value to the final score for the alignment. In order to allow the computer to create gaps in a biologically significant manner, the value contributed is a negative one—thus introducing a gap into an alignment is penalised. An initial gap is generally penalised by using the 'gap open penalty' and forms a relatively high cost to the program. Any subsequent gaps belonging to that particular insertion event are penalised at a much lower cost by 'gap extension penalties'. Such a scoring system, known as 'affine gap penalties' encourages longer gaps in the alignment instead of lots of smaller, more random gaps and results in a more biologically significant alignment.

Rigorous alignment software uses values for base or residue comparison, along with the penalties for gaps to create its own matrix. Each position in the matrix is occupied by the highest score possible between pairs of bases or residues based on the alignment up to that point. It can allow for gaps in the alignment and in such cases, points are deducted from each score to compensate for introducing them. Once the matrix has been filled, the computer can trace back the best alignment following the route of the highest values—the sum of which is returned with the alignment. In order to create the matrix, the most rigorous programs need memory capacity proportional to the multiple of the lengths of the two sequences to be aligned. This is approximately 1.6 KB for two sequences of 100 bases or residues in length, rising to 16 MB for sequences of 1000 bases or residues in length.

Depending on the requirements of the user, an alignment may either be local or global. Local alignments match only conserved regions within two given sequences and discard any unmatched portions so they do not appear in the final result. In computational terms, once the score goes down to such a point that no further alignment will increase it, the match is terminated and the matched portion brought back as a result (hit). Flexible local alignment programs will bring back several matches in a sequence, regardless of their topology. The most rigorous algorithm for local alignments is Smith–Waterman.

Global alignment programs are designed to match sequences end-to-end and will search for the best score possible. As they cannot discard any unmatching regions of sequence, it is best to use global alignment programs to align sequences of approximately the same length—for example orthologous genes or proteins. The most rigorous algorithm for local alignments is Needleman–Wunsch, which should be used for short, divergent sequences.

All programs designed to offer pairwise alignment will bring back a match, however bad, that represents the highest score possible. For proteins, if the subsequence alignment displays 40% identity, sequences are generally expected to be homologous—i.e. evolutionarily related. Should this identity fall below 20%, statistics suggest there can be no definition of homology based simply on alignment. The 'twilight zone' indicates the region in-between these two matches where identity of the alignment is not deemed to be statistically significant enough to indicate any homology and a relationship must be demonstrated using other methods.

Other, faster, alignment methods still use dynamic programming, but have cut corners in order to both speed up the alignment and reduce compute power. These are generally slightly less accurate than the most rigorous implementations, and should be used on sequences that are relatively closely related, or much longer sequences. These include pairwise alignment matches such as LAlign or, in more extreme cases, sequence search software such as BLAST or FastA. In addition, multiple sequence alignment options generally rely on initial pairwise alignment before producing a multiple match. Although extensions to the Smith–Waterman and Needleman–Wunsch algorithms can and have been made to simultaneously align three sequences, the compute power required is extraordinary and currently multiple alignment methods using other means offer a good solution at a much lower computational cost.

Multiple Sequence Alignment

Multiple sequence alignment (MSA) methods and their main applications. It is focused on progress made over the past decade. The three first sections review recent algorithmic developments for protein, RNA/DNA and genomic alignments. The fourth section deals with benchmarks and explores the relationship between empirical and simulated data, along with the impact on method developments. The last part of the review gives an overview on available MSA local reliability estimators and their dependence on various algorithmic properties of available methods.

Multiple sequence alignment (MSA) methods refer to a series of algorithmic solution for the alignment of evolutionarily related sequences, while taking into account evolutionary events such as mutations, insertions, deletions and rearrangements under certain conditions. These methods can be applied to DNA, RNA or protein sequences. Indeed, a large number of *in silico* analyses depend on MSA methods. These include domain analysis, phylogenetic reconstruction, motif finding and a whole range of other applications.

MSA is indeed an important modeling tool whose development has required addressing a complex combination of computational and biological problems. The computation of an accurate MSA has long been known to be an NP-complete problem, a situation that explains why over 100 alternative methods have been developed these past three decades.

Algorithmic Frameworks for MSA Computation

Despite their wide diversity, MSAMs all share a major key property: their reliance on approximate and usually greedy heuristics, imposed by the NP-complete nature of the problem. These heuristics all depend, more or less explicitly, on specific data properties, such as size, nature of the homology, relatedness, length and so on. As a consequence, any change—even minor—on the kind of data being modeled requires the development of novel heuristic strategies. Such changes have recently included the need of up scaling under the high-throughput sequencing pressure and the need for more complex sequence descriptors, including non-coding RNA or non-transcribed genomic sequences. Shifting modeling needs can also drive the developments of novel heuristics, a fact well illustrated by the recent development of phylogeny-aware aligners. Another driving force behind the development of new heuristics has been the increasing availability of structural data that has fueled the development of hybrid methods able to simultaneously deal with sequences and secondary (RNA) or tertiary (RNA and proteins) structures. Likewise, the explosion of available genomic data has put a lot of pressure on the development of a new generation of non-coding/ non-transcribed DNA aligners.

Commonly used Algorithms

Given a set of biological sequences (RNA, proteins, DNA), the purpose of a MSA method is to align the sequences in a way that will either reflect their evolutionary, functional or structural relationship. This purpose is achieved by inserting gaps of varying length within the sequences, allowing homologous positions to be aligned with each other—just like one would align beads of identical color in an abacus. In an evolutionary context, these gaps represent insertions and deletions (indels) within the genome that are hypothesized to have occurred during the evolution from a common ancestor. In a correct MSA, aligned residues should be maximally similar according to some specified criteria. For instance, if the alignment is meant for evolutionary reconstruction, the residues should be homologous, that is to say, correspond to the same residue in the last unique common ancestor of the considered sequences. If the alignment is meant to be a structural model, aligned residues should have comparable positions in their respective 2D or 3D structures. If the alignment is functional, as may happen when analyzing genomic data, aligned positions are expected to support similar functions. Even though it is reasonable to expect a significant overlap between these criteria, it must be stressed that the complexity of evolutionary forces is such that their full agreement cannot be taken for granted. For instance, two structures may be similar as a consequence of convergent evolution but non-homologous from an evolutionary point of view.

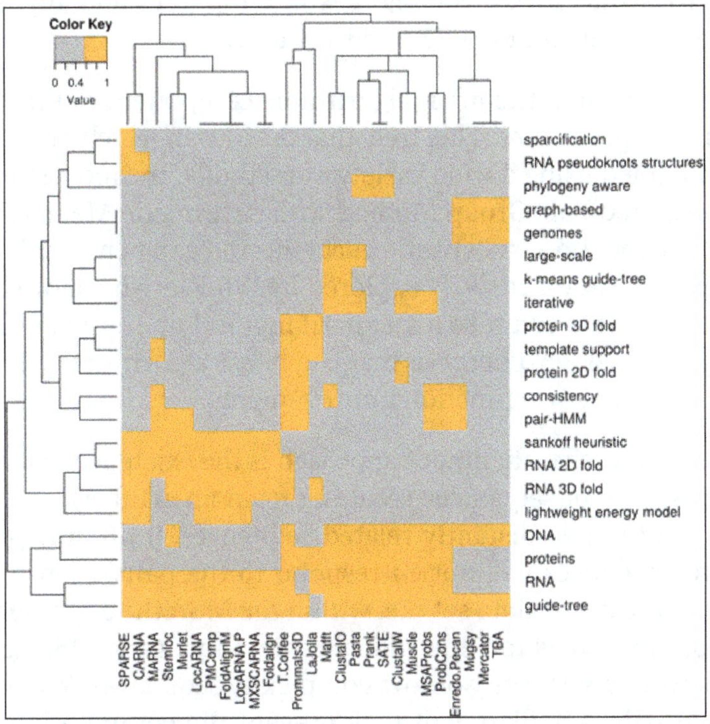

Main algorithmic components of the most widely used multiple aligners. On the heatmap, orange entries indicate a feature implemented in the considered method. Both the aligners and the components were clustered by similarity using the R-package.

To build an MSA, one needs a scoring function (objective function) able to quantify the relative merits of any alternative alignment with respect to the modeled relationship. The MSA can then be estimated by computing an optimally scoring model. The objective function is a critical parameter, as it precisely defines the modeling accuracy of an MSA and its predictive capacity. When it comes to evolutionary reconstructions, the most commonly used objective functions involve maximizing weighted similarities (as provided by a PAM or BLOSUM substitution matrix) while using an affine gap penalty to estimate indels costs. The substitution cost can be adjusted using tree-based weighting schemes that reflect the independent information contribution of each sequence, and the score of columns is estimated by considering the total all-against-all (sums-of-pairs) substitution cost. It is well known that the sum-of-pairs functions are unlikely to be modeling biological relationships accurately enough, but they have been shown to provide a reasonable trade-off between structural correctness and computability, that is to say, the possibility to rapidly estimate a reasonable MSA.

Under their most common formulations, the optimization of sums-of-pairs evaluation schemes is NP-complete. One therefore needs to rely on heuristics, the most common one being the progressive alignment algorithm initially described by Hogeweg and Hesper. This algorithm involves incorporating the input sequences one by one into the final model, following an inclusion order defined by a pre-computed guide tree. At each node, a pairwise alignment is carried out between either a pair of sequences, a sequence and a profile or two profiles. The pairwise alignments taking place at each node are estimated using more or less sophisticated adaptations of the Needlman and Wunsch global dynamic programming alignment algorithm. The combination between a tree-based progressive strategy and a global pairwise alignment algorithm forms the backbone of most available methods, including ClustalW, T-Coffeeand ProbCons. It is also particularly well adapted for the design of iterative strategies), involving reestimating trees and alignments until both converge, as implemented in MUSCLE, MAFFT and Clustal Omega.

Aside from the objective function, the main algorithmic component of the progressive alignment is the guide tree estimation procedure. This tree, that decides in which order the sequences will be incorporated, can be obtained using a wide variety of methods, the most standard being Neighbor Joining (NJ) and Unweighted Pair Group Method with Arithmetic Mean (UPGMA). The interaction between the objective function (substitution scheme and gap penalties), the weighting scheme and the tree is complex and was extensively explored by Wheelerwho showed how the proper tuning of these various components can take a standard method up to the level of the most accurate ones. It is therefore unsurprising to observe that the latest algorithmic developments have been focused on guide trees and objective function improvements.

The main caveat of the progressive alignment approach is the existence of local minima (high level of similarity between a subset of sequences resulting from an artifact). For instance, if the guide tree induces the alignment of two distantly related sequences, it often happens that the optimal alignment of these two sequences will not correspond to the pairwise projection one would get from the optimal MSA of the entire data set (i.e. within the MSA the alignment of the two sequences will be slightly suboptimal so as to allow global optimality at the MSA level). This situation is common when dealing with low-identity or low-complexity sequences. When this occurs, the early computation of the first pairwise alignment may prevent the computation of a globally optimal MSA.

The most common strategy to avoid local minima during a progressive alignment is the use of consistency, as originally described by. The rationale of consistency is relatively straightforward: given a set of sequences and their associated pairwise alignments, treated as constraints, scores for matching pairs of residues are reestimated so as to deliver pairwise alignments more likely to be compatible with a globally optimal MSA. The first strategy involving such a reestimation of match costs was reported by Morgenstern as overlapping weights. This scheme later inspired the T-Coffee scoring scheme that has become the archetypical progressive consistency-based aligner. Optimizing an alignment against a set of predefined constraint is known as the Maximum Weight Trace problem. It is NP-complete under its most common formulations and can only be solved for small instances. The T-Coffee algorithm is a heuristic approach that involves reestimating the initial costs of every potential pairwise match by taking into account its compatibility with the rest of the pairwise alignment. The resulting scoring scheme makes it more likely to assemble consistent sub-alignments during the progressive MSA procedure. The main strength of this approach is to allow the computation of MSAs even when an objective function is only available to be optimized at the pairwise level. Since then, the consistency-based approach has become one of the most popular algorithmic frameworks for the development of novel methods.

In a consistency-based algorithm, the most critical parameter is the primary library. Given a set of sequences, the primary library is a collection of all possible pairwise sequence comparisons.

This library is used to define the consistency-based objective function. In the original T-Coffee, the library was a compilation of all pairs of residues found aligned in the entire pairwise local and global alignments. These residue pairs were weighted according to the estimated reliability of their source alignments. Later on, a variation of the T-Coffee algorithm named ProbCons established the superiority of pair-HMM-based libraries. In ProbCons, libraries are compiled using a pair-HMM to estimate the posterior probability of all possible pairs of residues (between distinct sequences) to be aligned. The use of a pair-HMM soon became popular among other alignment methods. The main novel features of ProbCons over T-Coffee were the use of a more formal probabilistic framework, thanks to the HMM and the implementation of a biphasic gap penalty when estimating pairwise alignments. Algorithms relying on a similar combination are often referred to as probabilistic consistency algorithms; they include the PECAN multiple genome aligner, which uses the Durbin forward only divide and conquer pairwise alignment and MSAProbs, which relies on a partition function to achieve more informative posterior probabilities when compiling the library. Some degree of consistency was also incorporated in the MAFFT 'linsi' algorithm.

When benchmarked on structure-based reference alignments, consistency-based aligners have long been shown to yield the most accurate MSAs. This accuracy comes, however, at a significant memory and CPU cost, with most implementations being cubic in CPU and quadratic in memory with the number of sequences. Three strategies have been proposed to address this problem. The simplest one involves faster library computation. For instance, FM-Coffee, the fast implementation of T-Coffee computes its library using three fast aligners, and eventually extracts the resulting pairwise projections. The high correlation between the various projections then makes it possible to band the consistency extension and significantly lower time and memory complexity at a near-quadratic level. Even though the resulting alignments are not as accurate as those obtained using the default procedure, they tend to be more accurate than those produced individually by the combined methods. The second strategy involves parallelization. Two such schemes have been recently published Cloud-Coffee and MSAProbs, both which involve parallelizing library computation and the relaxation step during which pairwise costs are reestimated when the progressive alignment assembly is taking place. The last step, which involves splitting computation according to the tree topology, is highly dependent on the guide tree symmetry, best performances being achieved with perfectly balanced guide trees. The third strategy is more sophisticated and involves tuning the library granularity by considering sequence segments rather than single residues. This implementation, available in SeqAn, is especially well suited to long closely related sequences, in which long identical segments can be identified.

Large-scale Multiple Aligners

Even in their most optimized forms, consistency-based methods cannot deal with more than a few hundred sequences. This limit is rather severe in a context where the explosion of genomic sequence availability has resulted in unprecedented large homologous families that can require aligning up to 1.5 million members (number of ABC transporters in Pfam) and soon many more. While the biological relevance of large MSAs can be questioned, recent analysis indicates that important results can be established from such large models, thus making the accurate and efficient building of large MSAs one of the current grand challenges of modern biology. Three methods are currently able to deal with such large data sets: the PartTree mode of MAFFT, Clustal Omega and PASTA, the newest version of SATé. These methods share a common characteristic: their reliance on a fast pre-clustering step (sub-quadratic in time) that makes it possible to rapidly determine the order in which sequences should be aligned.

In the original progressive methods, the guide tree was estimated by comparing all the sequences against one another to estimate a distance matrix. This comparison can be based on a slow Needleman and Wunsch alignment or on a fast *k*-tuple vector comparison as implemented in MAFFT, MUSCLE and T-Coffee. The fast comparison does not, however, solve the issue of quadratic time and space requirements for the matrix computation followed by the cubic time complexity of tree estimation when using either UPGMA or NJ. These requirements become prohibitive when processing over 10 000 sequences. Recent clustering methods have been designed to address this issue. In Clustal Omega, the guide tree is estimated using the mBed method. The principle of mBed is to first estimate the distance between each sequence and a tiny subset of sequences selected on the basis of their length. For each sequence, the result is a distance vector that can be used to run a hierarchical *k*-means clustering, whose relatively low complexity (*NlogN* under the most common heuristic implementations) allows large data sets of 10 000 sequences or more to be aligned. PartTree relies on a slightly different procedure that also involves using a small set of seed sequences to rapidly pre-cluster the sequences. In both mBed and PartTree, the pre-cluster step is followed by the computation of sub-trees that are eventually combined together to form the guide tree. The PartTree approach was recently improved in the SATé algorithm, which involves an extra iterative tree-refinement step. The latest attempt at aligning large data sets is an adapted version of the T-Coffee algorithm that involves combining *k*-means clustering with consistency-based MSAs at a lower level. These approaches scale well, but at the cost of significantly lower accuracies when aligning >1000 sequences, as shown in the Clustal Omega benchmark analysis. A probable side effect of this decreased accuracy has been the report of high alignment inconsistencies between MAFFT, Clustal Omega and T-Coffee when dealing with large data sets of relatively similar orthologous mitochondrial sequences.

Phylogeny-aware Multiple Aligners

A major milestone in the development of MSAMs has been the introduction of structure-based reference alignments that can be used to compare the relative capacities of various methods to reconstruct structurally correct alignments from sequence only. The choice of structure seems rather natural because 3D features are known to be more evolutionary resilient than the underlying sequences. On the other hand, this approach relies on the unproven rationale that structurally and evolutionary correct alignments are identical. No proof exists that this assumption may be correct, and a simple reasoning suggests it may not be the case. Indeed, while there can be only one correct way of matching homologous residues—the one that perfectly reflects the unique evolutionary history of the considered sequences and matches—there can be as many structurally correct alignments as there are ways to superpose the sequences with equivalent 3D compactness. Another major potential discrepancy between structural and evolutionary alignments results from convergent evolution. Whenever such a process has shaped some portions of a sequence data set, the resulting alignment matching convergent regions will be structurally correct and evolutionary false—and reciprocally.

This issue has recently been addressed by a series of works aiming at evaluating multiple aligners' accuracy on the basis of the quality of the phylogenetic models they support. These aligners are referred to as phylogeny-aware aligners. PRANK was one of the first. It relies on the idea that correct MSAs must have indels patterns properly reflecting the underlying phylogenetic tree. PRANK was rapidly followed by SATé, an iterative multiple aligner derived from MAFFT that attempts to estimate the MSA supporting the highest-scoring maximum likelihood tree. An important merit of this approach is to depart from the long-held assumption that the best MSA is the one maximizing similarity between sequences. In the context of phylogeny-aware aligners, the best MSA is defined as the one yielding the best phylogenetic model. The consequences on the resulting alignments are rather significant and were particularly well illustrated in a recent analysis by Blackburne and Whelan who found that 'similarity-based' MSAMs (e.g. ClustalW, MUSCLE, ProbCons, MAFFT, T-Coffee) and 'evolution-based' MSAMs (e.g. PRANK and BAliPhy) tend to form discrete clusters under the multidimensional scaling based on their own similarity measures between pairs of alternative MSAs. The selected aligners have a substantial impact on downstream phylogenetic inference and report the tree topologies and branch length to depend on the aligner category. Aligners also have a clear impact when quantifying positive selection, with different readouts associated with various aligners as reported on the analysis of several *Drosophila* genomes. Morrison suggests that phylogeneticists are usually dissatisfied with similarity-based alignments and tend to manually edit their MSAs to produce alignments more likely to reflect homology from a true evolutionary stand-point. This observation may also explain why results and method ranking achieved on evolutionarily simulated data sets significantly differ from those measured on structure-based empirical data The same reliability index can be used to select both the most phylogenetically informative positions and the positions most likely to contain structurally analogous residues. It is worth mentioning that this study gave contrasting results with respect to phylogeny-aware aligners, and while SATé appears to perform well both on structure and evolutionary benchmarks, PRANK was found to return poor structural alignments, while being able to produce alignments supporting trees with an accuracy comparable with the other MSAMs—phylogeny aware and similarity based.

Structure-based MSAs

Thanks to its high evolutionary resilience, structural information can help produce high-quality models, especially in situations where one aims at modeling structural and functional relationships.

Protein Sequence/Structure Multiple Alignments using Template-based Protein Aligners

Structural information has long been known to be more resilient than its underlying sequence counterpart. Yet, it is only recently that the corpus of available structural information has made it worthy to develop methods able to combine sequence and structural information within a single model. While the first generation of methods used to rely on protein structure threading and related methods, the newer generation of aligners takes advantage of the availability of multiple experimental structures within an increasing number of protein families. It has become common practice to combine structural aligners output (pairwise or multiple) using a consistency-based framework. The principle is fairly straightforward and involves associating each sequence with a template that can be either a bona-fide structure or a sequence with a known structure related closely enough to the sequence of interest so that no ambiguity exists in the template/target sequence alignment. When no structure is available, sequences are replaced with profiles from which one can either generate a conservation profile (as done in PSI-Coffee or TM-Coffee) or a secondary structure prediction using PSIPRED, or both as done in Prommals3D, 3D-Coffee and Expresso. The library is then built by aligning the sequences in pairs, using the pairwise method best suited for the considered templates. In this way, alternative methods can be combined seamlessly. This approach is especially convenient when dealing with pairwise structural alignment methods lacking a multiple alignment implementation. The possibility of combining several alternative structural aligners also provides a simple way to address the difficulty of objectively telling alternative structure-based sequence alignment models apart. In this context, the consistency-based approach makes it possible to identify the portion of a model best supported by all the considered methods. This approach has been implemented in the Expresso package, which supports three of the most commonly used structural aligners and can easily accommodate any other third-party aligner.

RNA Multiple Sequence Aligners

The low-complexity alphabet of RNA molecules makes their alignment more challenging than that of protein sequences, with biologically meaningful alignments difficult to estimate <60% identity. Whenever secondary structures are evolutionarily conserved, covariation often becomes the strongest available signal. However, standard aligners, like ClustalW, MAFFT or T-Coffee, assume site independence and cannot take this information into account, at least in their default usage. In fact, for these standard aligners, covariation is more of a confounding factor as it decreases sequence identity. More specialized aligners are therefore needed, able to simultaneously recognize similarity at the sequence and secondary-structure level. These algorithms are all heuristic approximations, more or less explicitly related to the Sankoff dynamic programming algorithm, which simultaneously folds and aligns RNAs at a prohibitive computational cost $O(N^{3m})$, with m being the number of sequences and N their length. Several banded implementations of the algorithm have been reported. These enforce restrictions on the size or shape of substructures; they can be pairwise aligners such as Consan, Dynalign, Stemloc and Foldalign or multiple aligners such as MXSCARNA, a progressive multiple aligner based on SCARNA, a pairwise alignment method based on fixed-length stem fragments defined by means of McCaskill's algorithm. Murlet is another such aligner that first estimates the base pairing and match probabilities before running the Sankoff algorithm with these probabilities to estimate the final alignment. In MARNA, the structural information is used for pairwise RNA comparisons before joining them into a MSA with T-Coffee. PMcomp is a method for progressive multiple alignments based on a McCaskill's algorithm to generate and then compare base pairing probability matrices, which enables lightweight computation. For this purpose, it uses a base pair-based energy model instead of the original loop-based energy model.

PMcomp simplifies Sankoff's model by predicting only a single consensus structure and has been an important source of inspiration of the development of many RNA aligners such as LocARNA, FoldAlignM and LocARNA-P that use additional heuristics to further restrict the folding space to be explored, thus resulting in an $O(n^4)$ time complexity. CARNA extends the PMcomp model to pseudoknot structures. RAF combined the ideas of and, resulting in a lightweight Sankoff-variant with sequence-based speed up. SPARSE is one of the newest algorithm in the Sankoff-style category. It is reliable. It runs in quadratic time thanks to its reliance on 'sparsified' prediction and RNA alignments based on their structure ensembles. Compared with LocARNA, SPARSE achieves similar alignment and better folding quality in significantly less time (speedup: 3.7). Another approach that StrAl implements is a scoring scheme that combines sequence similarity with pairing probability. This fast heuristic allows a runtime similar to ClustalW. T-Lara implements a graph-based representation of sequence-structure alignments modeled using integer linear programming. The resulting alignments are then further integrated into a T-Coffee style library using Lagrangian relaxation and eventually resolved into an MSA model using T-Coffee. RNAsampler is a sampling-based algorithm able to find common RNA structures in multiple RNA sequences. The program probabilistically samples aligned RNA stems based on inter-sequence base alignment probabilities and stem conservation calculated from intra-sequence base-pairing probabilities. Another example is RNAcast, which for each sequence predicts structure profiles within a defined minimum free-energy threshold and then computes the optimal consensus structure that is shared by all the RNAs.

More recently, following up on T-Lara, systematic attempts were made to apply the consistency paradigm to secondary structure predictions. One can do so by considering libraries made of pairs of pairing residues. This principle has been developed in R-Coffee, which adopts a pre-folding approach, predicting with RNAplfold the shape of the individual RNA sequences in an early step. Subsequently, the program estimates the MSA with the highest agreement between structures and sequences. A similar approach was later developed in the RNA compliant version of MAFFT, where consistency is measured by combining pairs of paired residues across combination of triplets. Both packages achieve comparable levels of accuracy, the main strength of R-Coffee being its capacity to combine complex pairwise RNA aligners like Consan into highly accurate multiple aligners.

The scarcity of RNA 3D information probably explains why so little attention has so far been given to the generation of accurate 3D structure-based multiple RNA alignments. The situation is slowly changing with several novel algorithms recently described to deal with this problem. Existing tools include pairwise aligners like ARTS, SARA, DIAL and R3D Align, and multiple ones like SARSA, LaJolla and SARA-Coffee. The heuristic nature of these algorithms tends to make them error prone, hence the importance of RNA-specific MSA editors. Many such tools are available (4SALE, CONSTRUCT, JPHYDIT, RALEE, SARSE) and able to dynamically display secondary and compensatory information while editing RNA MSAs.

It is important to note that these algorithms only work well when dealing with RNA-containing evolutionary conserved secondary structure. In their validation of the SARA-Coffee algorithm, Kemena *et al.* reported that when <70% of the nucleotides are involved in evolutionary-conserved Watson and Crick base pairs, structure-aware aligners like those listed above tend to degrade alignment accuracy. This degradation is a mechanical consequence of the explicit algorithmic attempt to seek and match secondary structures under the assumption that these should be homologous. In practice, however, the structures may not be conserved, or not properly predicted, especially in cases where protein/RNA interaction play a role on the *in vivo* RNA fold. This problem is especially important when considering the issue of aligning long non-coding RNA (lncRNA), the most recently described class of RNA genes. So far, no indication of extensively conserved secondary structure has been reported for these genes, which makes it increasingly likely that this new category of transcripts will require a new generation of aligners in the years to come, possibly motif biased and drawing on the recent report that dinucleotide information can help improve lncRNA alignments.

Multiply Aligning Non-transcribed Sequences

The increasing availability of complete genomes makes it a pressing need to develop non-transcribed intergenic sequence alignment tools. Indeed, these sequences come with challenges of their own: extreme length, poor conservation, order variations (inversions, translocations and duplications) and the extreme molecular clock heterogeneity resulting from the wide range of functions supported in different ways by the untranslated part of the genome. This last issue is likely to become increasingly important as novel genomic functions, often associated with epigenetics.

Multiple Genome Alignments

While standard sequence aligners usually imply the modeling of three evolutionary operations, insertion, deletion and substitutions, genome-scale alignments must incorporate at least three more operations: inversions, translocations and duplications. In general, multiple genome aligners achieve this through two separate steps. In a first step, homologous genomic fragments are sorted into bins, and in a second step, these bins are turned into standard MSA models. This last step usually depends on standard progressive aligners.

While the first alignment step could rely on a simple clustering/segmentation approach, such a procedure would yield disconnected MSA blocks, giving few insights into genomic evolution. For this reason, most new-generation genome aligners rely on the sorting by reversal algorithm for the segmentation step. Sorting by reversal is an NP-complete problem that amounts to reconstructing the minimum chain of events that would edit one genome into another using a series of translocations and inversions. It is not necessary to solve this problem to align genomes, but it helps quantifying the evolutionary cost of alternative alignments. In practice, most algorithms start by seeking colinear segments, often relying on anchor points (usually proteins) gathered using an all-against-all BLAST procedure. The most popular procedures include Mercator that uses protein anchors, MUMS/Mems (Mugsy or the systematic use of local alignments.

TBA was one of the first algorithm to consider a multiple genome alignment (MGA) as a set of separate blocks rather than a continuous sequence, thus making data processing a necessary prerequisite). In the newest generation of MGAs, the pre-processing has become tightly integrated with the alignment process, as in Mercator-Mavid or Enredo/Pecan, which uses graph structures to identify the different genome rearrangements, splits the multiple genomes accordingly and feeds the resulting bins of multiple sequences to Pecan, a space-efficient consistency-based aligner using the Durbin forward-only linear space dynamic programming procedure. Other graph structures (e.g. A-Brujin graph, Cactus graph) have been used for this purpose (see Kehr *et al.* for a comparison of different graph structures). Another alternative is to simultaneously carry out alignment and segmentation in a progressive way. This procedure developed by Brudno uses the equivalent of consistency to identify rearrangements most likely to be supported by the whole data set.

MGA method development has, however, been hampered by the difficulty to objectively assess the relative merits of each aligner. In contrast with proteins or RNA sequences, no such thing as a structure or its equivalent is available for genomes, and when the Alignathon contest proposed to compare the capacities of MGAs on eukaryotic data, the benchmarking was eventually carried out using the PSAR objective function, a sequence-based estimator relying on probabilistic sampling. The PSAR objective function was initially developed to evaluate genomic MSAs. Its principle is somehow similar to the consistency-based approach of T-Coffee, though more complete and more computationally demanding. In PSAR, given a data set, all sequences are removed in turn, the remaining sequences realigned and the removed sequence realigned to the sub-alignment. The stability of the realignment with respect to the input MSA is then used to estimate the reliability of each residue positioning within the final alignment model. This procedure is generic with no constraint limiting it to nucleotide alignments. It has, however, so far only been tested and benchmarked on simulated genomic data sets.

The Alignathon contest remains the only generic attempt to compare the reliability of multiple genome aligners. 13 MGA packages were compared on either *Drosophila* genomes or artificially generated mammalian genomes. The Alignathon coordinators took the decision to use the PSAR objective function as a standard of truth. Such a decision comes with important caveat, possibly reflected in the clear dominance of PSAR-align—a package explicitly optimizing this function—over most alternative aligners. In the corresponding Jaccard index analysis, they found that on the non-simulated fly genomes, over 50% of the aligned positions—at the nucleotide levels—are inconsistent between pair of methods. Such dispersion should be taken as a measure of the complexity one faces when trying to develop a generic DNA aligner. In contrast, more focused effort on well-defined genomic regions can be used to deliver high-quality alignments of functionally homologous regions. This approach has been successfully developed and used to study eukaryotic genome promoters.

Multiple Promoter Alignments

MGAs aim at using the genome reordering information so as to better understand evolutionary relationships and possibly identify functional constraints associated with gene organization conservation. In this context, promoter multiple comparisons are probably the best example of functional multiple alignments, aiming at uncovering common regulatory patterns between related sequences. These patterns are used to reveal transcription factor binding sites (TFBS). From an algorithmic point of view, the problem can be separated into two distinct categories: motif discovery among unaligned non-homologous (or distantly related sequences) and regular MSAs.

Methods for the discovery and comparison of homologous promoter regions are more recent. They were initially reported for the discovery of TFBS, through a process often referred to as evolutionary foot printing. Several methods have been described for that purpose. For instance, in, potential binding sites are first predicted on single sequences and then used as anchors during the alignment process. Another strategy is to use an alternative scoring scheme on the positions within a sequence known to fit a regulatory element. This more or less amounts to dressing up a sequence with profile weight matrices that define a position-specific scoring scheme. The main limitation, however, of these motif-based methods is their reliance on pre-computed sets of reference motifs. As an alternative, one can simultaneously identify the motifs and align the sequences as proposed in. Other methods can also model inversions and translocation, thus taking into account the fast motif turnover reported in promoter regions. All these methods are computationally too intensive to scale-up over a few (usually two) sequences, and scalable alternatives have been proposed for multiple sequence analysis. It is also possible to fine-tune existing methods for multiple promoter alignments, as shown by Erb *et al.*

The tuning also tooks into account the discriminative capacity between alignments of orthologous and paralogous gene regions.

Benchmarking Multiple Aligners Accuracy

Quantifying the accuracy of multiple aligners is just as critical as aligning sequences, especially when considering the aligners approximate nature. This seemingly obvious aspect has been generally overlooked by the community as reflected by the relative lack of correlation between the packages overall usage and their reported accuracy. ClustalW, for instance—whose 42 000 citations suggest a global usage level higher than all other packages put together—has not been consistently reported as the most accurate method. This surprising observation probably reflects on a combination of factors. The most obvious is the relationship between benchmarks rankings and day-to-day usability. It is likely that ClustalW, even though it does not rank #1 on all benchmarks, is nonetheless sufficiently accurate for many modeling activities, especially when dealing with orthologous data sets. One may also speculate on the existence of a strong methodological inertia within the biological community, where tool usage tends to snowball through protocol recycling.

The most critical component of an MSA is its scoring/objective function, the mathematical formula that quantifies the total score and therefore defines optimality, given a set of sequences. The rest of the algorithm is an optimization procedure attempting to generate an MSA model that maximizes the objective function. It is well established that even the best objective functions are merely approximations trying to model the behavior of biological sequences. As a consequence, there is no guarantee that a perfectly optimized MSA will systematically result in the most biologically meaningful MSA. This is the reason why multiple aligners also need to be evaluated/benchmarked for their capacity to produce correct alignments. A benchmarking procedure relies on existing collections of reference alignments considered as gold standards. These reference MSAs are routinely used as predictors for the accuracy of a given aligner on a given type of data sets and have had a major influence on methodological developments. The latter three categories meet the criterion of reference data sets, in that they can be pre-compiled and used to quantify the relative merits of one aligner over another. The simulation-based benchmarks, however, define an objective function rather than a benchmark procedure and cannot be considered a benchmark measure in the same sense as the others.

Structure-based Protein Benchmarks

Multiple aligners benchmarking has largely been driven by the use of structure-based reference MSAs, BAliBASE being the most widely used. These benchmarks all rely on structure-based reference alignments to evaluate any aligner able to handle their sequences. It has become customary to report new aligners along with the benchmark readouts established on at least two available structure-based reference data sets. This probably owes to the surprisingly low consistency between benchmarks. Indeed, as shown in, aligner's rankings established on the basis of the most common benchmarks are on average <50% consistent. This means that if a given benchmark suggests that method *A* is more accurate than method *B*, there is less than one chance of two that the same ranking is supported by another benchmark collection. In his detailed analysis of available benchmarks, Edgar suggested SABmark to be the most complete and informative, but only when using a subset of SABmark made of compatible pairwise structural alignments.

The growing need for large-scale aligners has resulted in the development of a new benchmark generation able to estimate alignment accuracy when assembling large data sets. The main issue when doing so is the scarcity of structural information. Of the 16 230 Pfam families with experimental structural information, about 50% merely have one member with a known 3D structure, and 25% have two members only. To accommodate this limitation, reference data sets were built by embedding sequences with a known structure within larger data sets made of sequences with unknown structure. This approach already used in PREFAB—with two sequences of known structure embedded within a data set of 50 sequences—has been extended in HomFam, so as to define much larger data sets of up to 100 000 sequences in which an average of 10 sequences with known structures are embedded. When doing so, accuracy is estimated by first aligning the large data sets. The projections of sequences with known structures are then extracted and accuracy is quantified by comparing these projections with the reference. In this procedure, the main caveat lies in the assumption that the seed sequence accuracy reflects well the global data set. This assumption is, however, only correct if the sequences with known structures are evenly distributed within the considered data set.

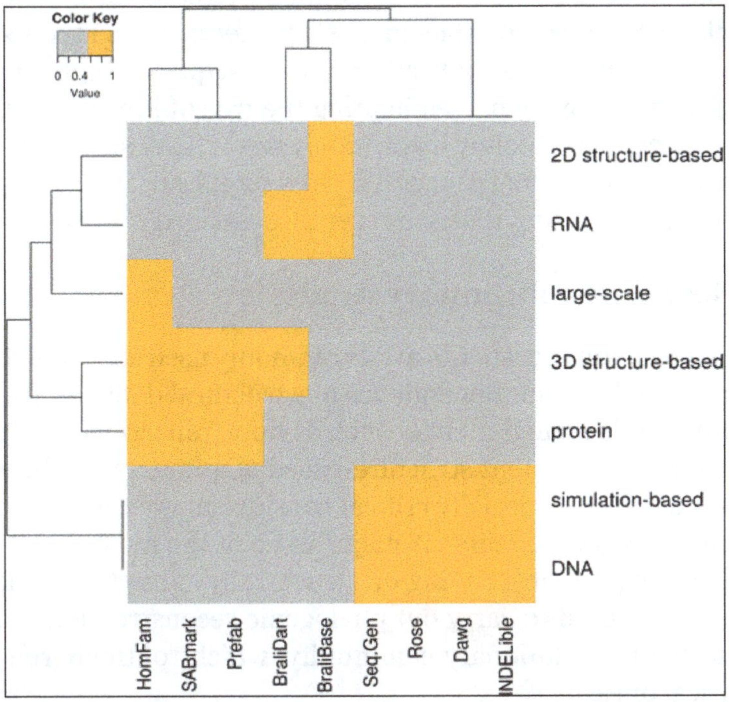

Main benchmark methods and their most relevant properties. On the heatmap, orange entries indicate a property describing a given method. Both properties and benchmarks were clustered by similarity.

Structure-based benchmarking does not necessarily depend on a reference alignment, and alternative methods have also been designed that rely on structural superposition rather than structural superposition-induced alignments. These developments were mostly the consequence of work by Lackner, who reported on situations where the structure-based superposition is ambiguous enough to support equally well several alternative sequence alignments. When this occurs, the reference alignment becomes the arbitrary prioritization of one reference over another, thus biasing the benchmark process. Most reference benchmarks deal with this problem by specifying core regions in which the reference alignment is expected to be less ambiguous, but this procedure remains dependent on the way in which core regions are defined. A more general alternative exists that involves comparing intra-molecular distances between pairs of aligned residue pairs. This measure, named iRMSD, makes it possible to quantify the structural fit implied by an alignment without having to rely on a reference.

Structure based RNA Alignment Benchmarks

Structural benchmarks have also been developed for RNA alignment evaluation. Three such benchmarks exist. BAliBASE is the most commonly used. It makes it possible to evaluate the accuracy of a multiple aligner on RNA sequences by considering the modeling capacity of the evaluated aligner with respect to some reference secondary structure. This dependence on sequence (on which the secondary structure estimation is based) slightly limits its scope, as it implies common dependencies between the reference compilation and the evaluation procedure. BraliDart, a newer data set, that is only based on structural information and contains sets of homologous RNA families with known experimental structures, has been recently reported. This data set is limited by the relative scarceness of experimental RNA 3D structures. Another specificity of BraliDart is its non-reliance on a reference structural alignment but rather on the structural fit implied by the sequence alignment using a distance RMSD measure, as defined by the iRMSD method. The third main category of RNA benchmark is made of ribosomal RNA reference alignments. They have not been assembled for benchmarking purposes, but rather as a consequence of the importance of accurate ribosomal RNA (rRNA) alignments when estimating the tree of life. These alignments have been done manually while taking into account highly conserved rRNA secondary structures that play critical roles in the ribosome functional capacities. No reference data set has yet been published to validate the MSAs of long non-coding RNAs, a recently described population of transcripts.

Simulated Data Sets for Evolutionary Analysis

Although empirical data benchmarks are the most commonly used strategies to evaluate alignment methods, they remain limited by their dependence on structural data and the lack of such data for the evaluation of certain kinds of alignments—such as non-transcribed DNA. Furthermore, it remains to be established to which extent structure-based alignments can be expected to be evolutionarily correct. This question is especially critical considering that phylogenetic modeling is one of the main applications of MSA modeling. A major issue of the most popular aligner methods is their systematic reliance, and possible tuning on structurally correct sequence alignments. These methods are, however, often used to carry out phylogenic reconstruction. This inconsistency has long been pointed out by the evolutionary community, which routinely relies on simulated data sets rather than empirical ones.

Simulated data sets rely on models mimicking evolution to generate sequences whose diversity is expected to represent a true evolutionary process. The main strength of this approach is to provide a perfectly traceable model, in which the relationship between nucleotides or amino acids is explicitly known. Their most obvious drawback is to rely on evolutionary models assumed to be correct, while the true extent to which they represent biologically realistic scenarios remains unknown. In any case, these approaches are useful when estimating the impact of extreme conditions on modeling capacity, for instance accelerated evolution, long-branch attraction and similar effects that can confound standard analysis. Several packages have been designed to generate simulated data sets, the most widely used being Rose, Seq-Gen, Dawg or INDELible.

When using these packages, the simulated alignments are considered as 'true' alignment, thus making it possible to use the same scoring system (Sum of Pairs Score, SP, or Column Score, CS) as for empirical benchmarks. It is worth noting that whenever simulated and structure-based reference data sets have been used to validate similar algorithms for alignment accuracy, the rankings were found to differ significantly between these two groups of benchmarks, a clear indication that different alignment characteristics are being evaluated. All phylogeny-aware aligners are currently evaluated using these simulated data sets. When doing so, the evaluation is often done on tree modeling capacity rather than on the MSA itself. Such algorithms include.

Resolving the apparent discrepancies between structure-based and simulated reference data sets will probably require a better understanding of the complex relation between alignment accuracy and trustworthy phylogenetic reconstruction. Moving one step in this direction, Dessimoz and Gil recently introduced tree-based tests of alignment accuracy, which not only use large and representative samples of real biological data, but also enable the evaluation of the effect of gap placement on phylogenetic inference. The use of empirical data sets obtained by enriching collections of orthologous genes in families likely to support the Tree of Life. When using such data sets, the discrepancy between phylogenetic and structural correctness appears to be less marked.

Quality Indexes for the Estimation of MSA Reliability

Given the approximate nature of all available aligners, identifying the trustworthy portions of an alignment is probably of higher practical importance than knowing the overall expected accuracy. Over the past few years, new methods have been reported aiming precisely at this. They can be roughly divided in three categories: those using structural information to assess protein or RNA accuracy, those depending on a conservation index to identify the positions most likely to be correct and methods depending on some form of local numerical instability to identify the most stable portions of an MSA model.

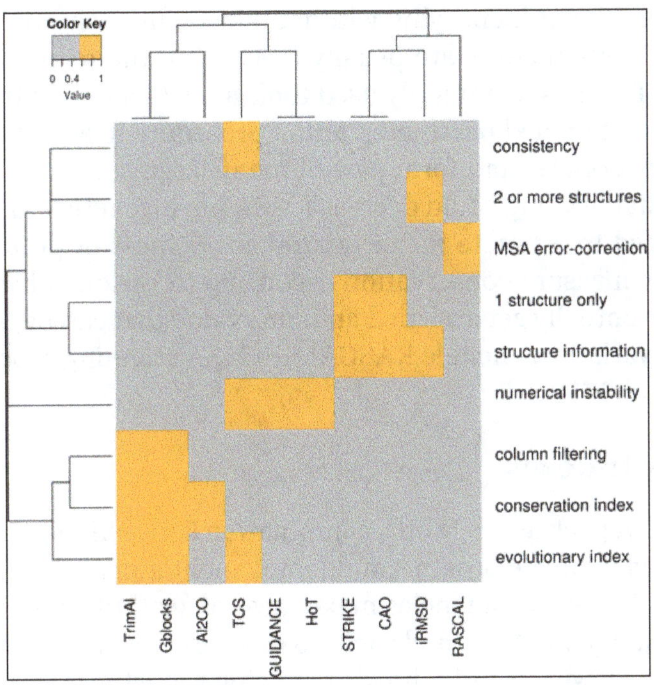

MSA quality indexes and their features. Features with zero are not used by the specific quality index.

Structural Conservation Indexes

With increasingly available structural data, the systematic use of 3D information for the monitoring of MSA accuracy is slowly becoming a realistic prospect. The first such methods were designed using the structural accuracy measured on all possible pairs of sequences with a known 3D structure as a proxy for global accuracy. This approach is useful but suffers from the major limitation that is the limited number of protein and RNA families for which more than one structure is available (about 25% of all PFAM families with a known structure, and <1% in RFAM). Recent efforts were therefore focused toward the use of single structures to estimate MSA accuracy. The CAO contact substitution matrix is one of the earliest work in this direction. The principle is to embed a sequence with a known structure in the MSA. This structure is then used to identify putative amino-acid contacts and the corresponding columns are then reevaluated using the CAO, a 400 × 400 substitution matrix assigning a score to every possible contact substitution. Unfortunately, the estimation of this matrix is limited by the lack of available data. This problem was addressed by the STRIKE algorithm, in which the contact substitution matrix is replaced with a contact potential metrics that considers the score of all potential contacts, as obtained from structural data. When using this matrix to evaluate an MSA, column contacts—as implied by at least one embedded structure—are evaluated by summing the contact score found in the contact log-odd matrix. This approach was shown to be significantly superior to CAO as a mean to discriminate between alternative alignments.

Sequence Conservation Indexes

Sequence conservation is one of the most straightforward ways of estimating MSA accuracy. A large number of tools have been developed for this purpose that roughly fall in two main categories: structural (i.e. structural estimates using sequence information) and evolutionary indexes. The evolutionary indexes aim at identifying within an MSA all positions likely to hamper phylogenetic reconstruction. These indexes are usually focused on the removal of diverse columns or indel-enriched regions. The most commonly used tools are Gblocks and trimAl, a re-implementation of Gblocks using an automated parameterization procedure to adjust the filtering level. While these tools are extremely popular and form part of many large-scale phylogenetic pipelines. Two recent reports suggest that filtering could decrease MSA phylogenetic modeling potential. Similar tools have been developed to estimate the structural correctness of protein MSAs. The simplest ones like AL2CO merely measure conservation according to various physicochemical criterions. Columns and residues eventually get assigned an index value that can be used when doing modeling. More sophisticated variations include RASCAL, an MSA scanning procedure meant to identify spurious regions within an MSA.

Alignment Stability Indexes

The most widely used MSA packages rely on a combination between the progressive algorithm and more or less sophisticated dynamic programming implementations, allowing pairwise alignments of sequences or profiles. These dependencies make these algorithms inherently unstable. Over the past few years, the development of methods able to quantify this instability to estimate local reliability has become a fast growing trend. The idea of using robustness as an indicator of biological accuracy is not new and had already been used as early as 1996 in a procedure that involved removing in turn every pair of amino acid in a pair of sequences before realigning them, so as to assess local alignment stability. Later on, the T-Coffee objective function was used to show the predictive power of consistency. In general, any procedure that may be used to perturbate an alignment lends itself to the definition of a robustness index. Such indexes can then be evaluated for their correlation with structural or phylogenetic modeling potential. The Head or Tail (HoT) procedure is a good example of a simple method (sequences are simply inverted), yielding useful information at the cost of a moderate computational overhead. Other similar procedure albeit more costly have been described. PSAR is one of them. It is a method that involves generating several alternative MSAs while removing each sequence in turn. Another such procedure is named GUIDANCE, where the MSA is reestimated several times using guide trees estimated from bootstrap replicates of the original MSA. The main issue with these two approaches is their relatively high computational cost. These methods are, however, much more informative than their sequence conservation alternatives. In a recent report, the T-Coffee consistency score has been shown to outperform both HoT and GUIDANCE for the identification of structurally correct portions within MSAs and both trimAL and Gblocks for the construction of accurate phylogenetic trees.

References

- Sequence-analysis, medicine-and-dentistry: sciencedirect.com, Retrieved 25, April 2020
- Virtual-Laboratory: bio.iitb.ac.in, Retrieved 16, January 2020
- Briefings-in-Bioinformatics-1745401: oup.com, Retrieved 21, August 2020

Chapter 3

Genome Analysis

The identification and comparison of genomic features such as DNA sequence, gene expression and structural variation is known as genome analysis. It falls under a branch of biology known as genomics. This chapter has been written carefully in order to provide a better understanding of all the topics related to genome analysis.

Genomics

Genomics refers to the study of the genome in contrast to genetics which refers to the study of genes and their roles in inheritance. Genomics can be considered a discipline in genetics. It applies recombinant DNA, DNA sequencing methods, and bioinformatics to sequence, assemble, and analyze the function and structure of genomes (the *complete* set of DNA within a single cell of an organism). Advances in genomics have triggered a revolution in discovery-based research to understand even the most complex biological systems such as the brain. The field includes efforts to determine the entire DNA sequence of organisms and fine-scale genetic mapping. The field also includes studies of intragenomic phenomena such as heterosis, epistasis, pleiotropy and other interactions between loci and alleles within the genome. In contrast, the investigation of the roles and functions of single genes is a primary focus of molecular biology or genetics and is a common topic of modern medical and biological research. Research carried out into single genes does not generally fall into the definition of genomics unless the aim of this genetic, pathway, and functional information analysis is to elucidate its effect on, place in, and response to the entire genomes networks.

History

Etymology

From the *gen*, "gene" (gamma, epsilon, nu, epsilon) meaning "become, cre-ate, creation, birth", and subsequent variants: genealogy, genesis, genetics, genic, genomere, genotype, genus etc. While the word *genome* (from the German *Genom*, attributed to Hans Winkler) was in use in English as early as 1926, the term *genomics* was coined by Tom Roderick, a geneticist at the Jackson Laboratory (Bar Harbor, Maine), over beer at a meeting held in Maryland on the mapping of the human genome in 1986.

Early Sequencing Efforts

Following Rosalind Franklin's confirmation of the helical structure of DNA, James D. Watson and Francis Crick's publication of the structure of DNA in 1953 and Fred Sanger's publication of the Amino acid sequence of insulin in 1955, nucleic acid sequencing became a major target of early molecular biologists. In 1964, Robert W. Holley and colleagues published the first nucleic acid sequence ever determined, the ribonucleotide sequence of alanine transfer RNA. Extending this work, Marshall Nirenberg and Philip Leder revealed the triplet nature of the genetic code and were able to determine the sequences of 54 out of 64 codons in their experiments. In 1972, Walter Fiers and his team at the Laboratory of Molecular Biology of the University of Ghent (Ghent, Belgium) were the first to determine the sequence of a gene: the gene for Bacteriophage MS2 coat protein. Fiers' group expanded on their MS2 coat protein work, determining the complete nucleotide-sequence of bacteriophage MS2-RNA (whose genome encodes just four genes in 3569 base pairs [bp]) and Simian virus 40 in 1976 and 1978, respectively.

DNA-sequencing Technology Developed

Walter Gilbert

Frederick Sanger

Frederick Sanger and Walter Gilbert shared half of the 1980 Nobel Prize in chemistry for independently developing methods for the sequencing of DNA.

In addition to his seminal work on the amino acid sequence of insulin, Frederick Sanger and his colleagues played a key role in the development of DNA sequencing techniques that enabled the establishment of comprehensive genome sequencing projects. In 1975, he and Alan Coulson published a sequencing procedure using DNA polymerase with radiolabelled nucleotides that he called the *Plus and Minus technique*. This involved two closely related methods that generated short oligonucleotides with defined 3' termini. These could be fractionated by electrophoresis on a polyacrylamide gel and visualised using autoradiography. The procedure could sequence up to 80 nucleotides in one go and was a big improvement, but was still very laborious. Nevertheless, in 1977 his group was able to sequence most of the 5,386 nucleotides of the single-stranded bacteriophage φX174, completing the first fully sequenced DNA-based genome. The refinement of the *Plus and Minus* method resulted in the chain-termination, or Sanger method, which formed the basis of the techniques of DNA sequencing, genome mapping, data storage, and bioinformatic analysis most widely used in the following quarter-century of research. In the same year Walter Gilbert and Allan Maxam of Harvard University independently developed the Maxam-Gilbert method (also known as the *chemical method*) of DNA sequencing, involving the preferential cleavage of DNA at known bases, a less efficient method. For their groundbreaking work in the sequencing of nucleic acids, Gilbert and Sanger shared half the 1980 Nobel Prize in chemistry with Paul Berg (recombinant DNA).

Complete Genomes

The advent of these technologies resulted in a rapid intensification in the scope and speed of completion of genome sequencing projects. The first complete genome sequence of an eukaryotic organelle, the human mitochondrion (16,568 bp, about 16.6 kb [kilobase]), was reported in 1981, and the first chloroplast genomes followed in 1986. In 1992, the first eukaryotic chromosome, chromosome III of brewer's yeast *Saccharomyces cerevisiae* (315 kb) was sequenced. The first free-living organism to be sequenced was that of *Haemophilus influenzae* (1.8 Mb [megabase]) in 1995. The following year a consortium of researchers from laboratories across North America, Europe, and Japan announced the completion of the first complete genome sequence of a eukaryote, *S. cerevisiae* (12.1 Mb), and since then genomes have continued being sequenced at an exponentially growing pace. As of October 2011, the complete sequences are available for: 2,719 viruses, 1,115 archaea and bacteria, and 36 eukaryotes, of which about half are fungi.

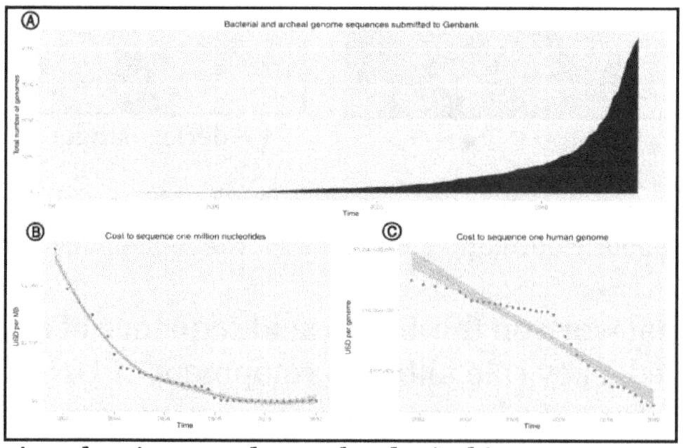

The number of genome projects has increased as technological improvements continue to lower the cost of sequencing. (A) Exponential growth of genome sequence databases since 1995. (B) The cost in US Dollars (USD) to sequence one million bases. (C) The cost in USD to sequence a 3,000 Mb (human-sized) genome on a log-transformed scale.

Most of the microorganisms whose genomes have been completely sequenced are problematic pathogens, such as *Haemophilus influenzae*, which has resulted in a pronounced bias in their phylogenetic distribution compared to the breadth of microbial diversity. Of the other sequenced species, most were chosen because they were well-studied model organisms or promised to become good models. Yeast (*Saccharomyces cerevisiae*) has long been an important model organism for the eukaryotic cell, while the fruit fly *Drosophila melanogaster* has been a very important tool (notably in early pre-molecular genetics). The worm *Caenorhabditis elegans* is an often used simple model for multicellular organisms. The zebrafish *Brachydanio rerio* is used for many developmental studies on the molecular level, and the flower *Arabidopsis thaliana* is a model organism for flowering plants. The Japanese pufferfish (*Takifugu rubripes*) and the spotted green pufferfish (*Tetraodon nigroviridis*) are interesting because of their small and compact genomes, which contain very little noncoding DNA compared to most species. The mammals dog (*Canis familiaris*), brown rat (*Rattus norvegicus*), mouse (*Mus musculus*), and chimpanzee (*Pan troglodytes*) are all important model animals in medical research.

A rough draft of the human genome was completed by the Human Genome Project in early 2001, creating much fanfare. This project, completed in 2003, sequenced the entire genome for one specific person, and by 2007 this sequence was declared "finished" (less than one error in 20,000 bases and all chromosomes assembled). In the years since then, the genomes of many other individuals have been sequenced, partly under the auspices of the 1000 Genomes Project, which announced the sequencing of 1,092 genomes in October 2012. Completion of this project was made possible by the development of dramatically more efficient sequencing technologies and required the commitment of significant bioinformatics resources from a large international collaboration. The continued analysis of human genomic data has profound political and social repercussions for human societies.

The "Omics" Revolution

The English-language neologism omics informally refers to a field of study in biology ending in *-omics*, such as genomics, proteomics or metabolomics. The related suffix -ome is used to address the objects of study of such fields, such as the genome, proteome or metabolome respectively. The suffix *-ome* as used in molecular biology refers to a *totality* of some sort; similarly omics has come to refer generally to the study of large, comprehensive biological data sets. While the growth in the use of the term has led some scientists (Jonathan Eisen, among others) to claim that it has been oversold, it reflects the change in orientation towards the quantitative analysis of complete or near-complete assortment of all the constituents of a system. In the study of symbioses, for example, researchers which were once limited to the study of a single gene product can now simultaneously compare the total complement of several types of biological molecules.

Genome Analysis

After an organism has been selected, genome projects involve three components: the sequencing of DNA, the assembly of that sequence to create a representation of the original chromosome, and the annotation and analysis of that representation.

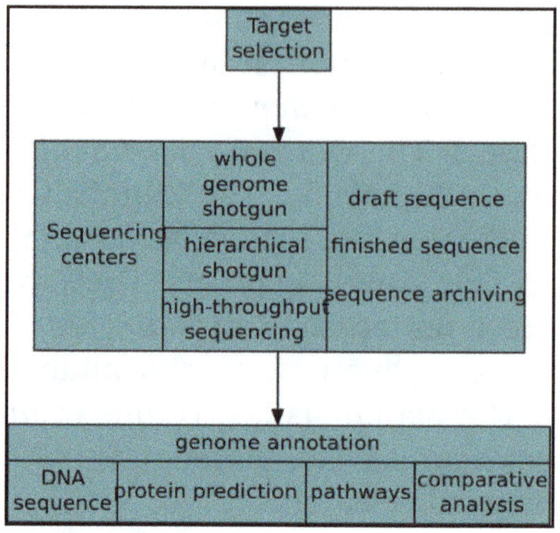

Overview of a genome project. First, the genome must be selected, which involves several factors including cost and relevance. Second, the sequence is generated and assembled at a given sequencing center (such as BGI or DOE JGI). Third, the genome sequence is annotated at several levels: DNA, protein, gene pathways, or comparatively.

Sequencing

Historically, sequencing was done in *sequencing centers*, centralized facilities (ranging from large independent institutions such as Joint Genome Institute which sequence dozens of terabases a year, to local molecular biology core facilities) which contain research laboratories with the costly instrumentation and technical support necessary. As sequencing technology continues to improve, however, a new generation of effective fast turnaround benchtop sequencers has come within reach of the average academic laboratory. On the whole, genome sequencing approaches fall into two broad categories, *shotgun* and *high-throughput* (aka *next-generation*) sequencing.

Shotgun Sequencing

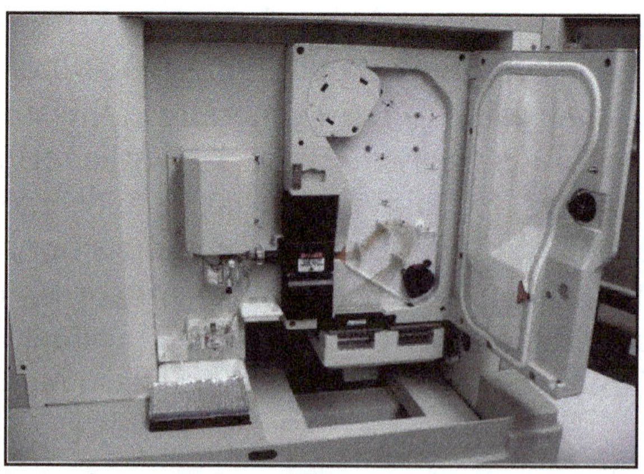

An ABI PRISM 3100 Genetic Analyzer. Such capillary sequencers automated early large-scale genome sequencing efforts.

Shotgun sequencing (Sanger sequencing is used interchangeably) is a sequencing method designed for analysis of DNA sequences longer than 1000 base pairs, up to and including entire chromosomes. It is named by analogy with the rapidly expanding, quasi-random firing pattern of a shotgun. Since the chain termination method of DNA sequencing can only be used for fairly short strands (100 to 1000 base pairs), longer DNA sequences must be broken into random small segments which are then sequenced to obtain *reads*. Multiple overlapping reads for the target DNA are obtained by performing several rounds of this fragmentation and sequencing. Computer programs then use the overlapping ends of different reads to assemble them into a continuous sequence. Shotgun sequencing is a random sampling process, requiring over-sampling to ensure a given nucleotide is represented in the reconstructed sequence; the average number of reads by which a genome is over-sampled is referred to as coverage.

For much of its history, the technology underlying shotgun sequencing was the classical chain-termination method, which is based on the selective incorporation of chain-terminating dideoxynucleotides by DNA polymerase during in vitro DNA replication. Developed by Frederick Sanger and colleagues in 1977, it was the most widely used sequencing method for approximately 25 years. More recently, Sanger sequencing has been supplanted by "Next-Gen" sequencing methods, especially for large-scale, automated genome analyses. However, the Sanger method remains in wide use in 2013, primarily for smaller-scale projects and for obtaining especially long contiguous DNA sequence reads (>500 nucleotides). Chain-termination methods require a single-stranded DNA template, a DNA primer, a DNA polymerase, normal deoxynucleosidetriphosphates (dNTPs), and modified nucleotides (dideoxyNTPs) that terminate DNA strand elongation. These chain-terminating nucleotides lack a 3'-OH group required for the formation of a phosphodiester bond between two nucleotides, causing DNA polymerase to cease extension of DNA when a ddNTP is incorporated. The ddNTPs may be radioactively or fluorescently labelled for detection in automated sequencing machines. Typically, these automated DNA-sequencing instruments (DNA sequencers) can sequence up to 96 DNA samples in a single batch (run) in up to 48 runs a day.

High-throughput Sequencing

The high demand for low-cost sequencing has driven the development of high-throughput sequencing (or next-generation sequencing [NGS]) technologies that parallelize the sequencing process, producing thousands or millions of sequences at once. High-throughput sequencing technologies are intended to lower the cost of DNA sequencing beyond what is possible with standard dye-terminator methods. In ultra-high-throughput sequencing as many as 500,000 sequencing-by-synthesis operations may be run in parallel.

Illumina (Solexa) Sequencing

Solexa, now part of Illumina, developed a sequencing method based on reversible dye-terminators technology acquired from Manteia Predictive Medicine in 2004. This technology had been invented and developed in late 1996 at Glaxo-Welcome's Geneva

Biomedical Research Institute (GBRI), by Dr. Pascal Mayer and Dr Laurent Farinelli. In this method, DNA molecules and primers are first attached on a slide and amplified with polymerase so that local clonal colonies, initially coined "DNA colonies", are formed. To determine the sequence, four types of reversible terminator bases (RT-bases) are added and non-incorporated nucleotides are washed away. Unlike pyrosequencing, the DNA chains are extended one nucleotide at a time and image acquisition can be performed at a delayed moment, allowing for very large arrays of DNA colonies to be captured by sequential images taken from a single camera.

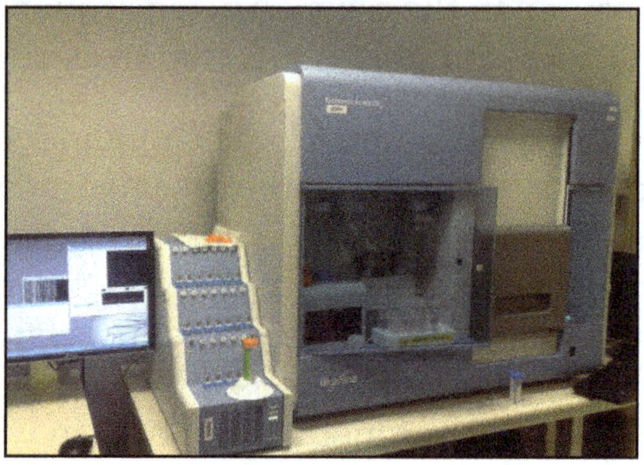

Illumina Genome Analyzer II System. Illumina technologies have set the standard for high throughput massively parallel sequencing.

Decoupling the enzymatic reaction and the image capture allows for optimal throughput and theoretically unlimited sequencing capacity. With an optimal configuration, the ultimately reachable instrument throughput is thus dictated solely by the analogic-to-digital conversion rate of the camera, multiplied by the number of cameras and divided by the number of pixels per DNA colony required for visualizing them optimally (approximately 10 pixels/colony). In 2012, with cameras operating at more than 10 MHz A/D conversion rates and available optics, fluidics and enzymatics, throughput can be multiples of 1 million nucleotides/second, corresponding roughly to 1 human genome equivalent at 1x coverage per hour per instrument, and 1 human genome re-sequenced (at approx. 30x) per day per instrument (equipped with a single camera). The camera takes images of the fluorescently labeled nucleotides, then the dye along with the terminal 3' blocker is chemically removed from the DNA, allowing the next cycle.

Ion Torrent

Ion Torrent Systems Inc. developed a sequencing approach based on standard DNA replication chemistry. This technology measures the release of a hydrogen ion each time a base is incorporated. A microwell containing template DNA is flooded with a single nucleotide, if the nucleotide is complementary to the template strand it will be incorporated and a hydrogen ion will be released. This release triggers an ISFET ion sensor. If a homopolymer is present in the template sequence multiple nucleotides will be incorporated in a single flood cycle, and the detected electrical signal will be proportionally higher.

Assembly

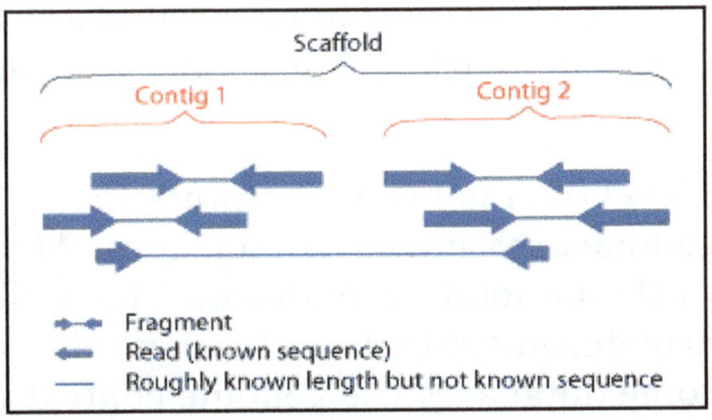

Overlapping reads form contigs; contigs and gaps of known length form scaffolds.

Multiple, fragmented sequence reads must be assembled together on the basis of their overlapping areas.

Sequence assembly refers to aligning and merging fragments of a much longer DNA sequence in order to reconstruct the original sequence. This is needed as current DNA sequencing technology cannot read whole genomes as a continuous sequence, but rather reads small pieces of between 20 and 1000 bases, depending on the technology used. Typically the short fragments, called reads, result from shotgun sequencing genomic DNA, or gene transcripts (ESTs).

Assembly Approaches

Assembly can be broadly categorized into two approaches: *de novo* assembly, for genomes which are not similar to any sequenced in the past, and comparative assembly, which uses the existing sequence of a closely related organism as a reference during assembly. Relative to comparative assembly, *de novo* assembly is computationally difficult (NP-hard), making it less favorable for short-read NGS technologies.

Finishing

Finished genomes are defined as having a single contiguous sequence with no ambiguities representing each replicon.

Annotation

The DNA sequence assembly alone is of little value without additional analysis. Genome annotation is the process of attaching biological information to sequences, and consists of three main steps:

1. identifying portions of the genome that do not code for proteins,
2. identifying elements on the genome, a process called gene prediction,
3. attaching biological information to these elements.

Automatic annotation tools try to perform these steps *in silico*, as opposed to manual annotation (a.k.a. curation) which involves human expertise and potential experimental verification. Ideally, these approaches co-exist and complement each other in the same annotation pipeline.

Traditionally, the basic level of annotation is using BLAST for finding similarities, and then annotating genomes based on homologues. More recently, additional information is added to the annotation platform. The additional information allows manual annotators to deconvolute discrepancies between genes that are given the same annotation. Some databases use genome context information, similarity scores, experimental data, and integrations of other resources to provide genome annotations through their Subsystems approach. Other databases (e.g. Ensembl) rely on both curated data sources as well as a range of software tools in their automated genome annotation pipeline. *Structural annotation* consists of the identification of genomic elements, primarily ORFs and their localisation, or gene structure. *Functional annotation* consists of attaching biological information to genomic elements.

Sequencing Pipelines and Databases

The need for reproducibility and efficient management of the large amount of data associated with genome projects mean that computational pipelines have important applications in genomics.

Research Areas

Functional Genomics

Functional genomics is a field of molecular biology that attempts to make use of the vast wealth of data produced by genomic projects (such as genome sequencing projects) to describe gene (and protein) functions and interactions. Functional genomics focuses on the dynamic aspects such as gene transcription, translation, and protein–protein interactions, as opposed to the static aspects of the genomic information such as DNA sequence or structures. Functional genomics attempts to answer questions about the function of DNA at the levels of genes, RNA transcripts, and protein products. A key characteristic of functional genomics studies is their genome-wide approach to these questions, generally involving high-throughput methods rather than a more traditional "gene-by-gene" approach.

A major branch of genomics is still concerned with sequencing the genomes of various organisms, but the knowledge of full genomes has created the possibility for the field of functional genomics, mainly concerned with patterns of gene expression during various conditions. The most important tools here are microarrays and bioinformatics.

Structural Genomics

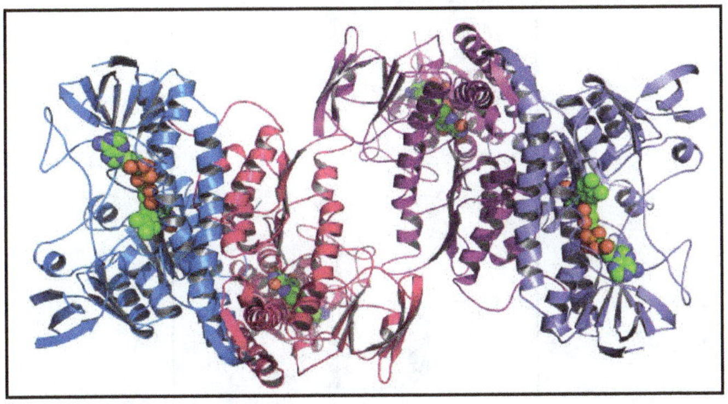

An example of a protein structure determined by the Midwest Center for Structural Genomics.

Structural genomics seeks to describe the 3-dimensional structure of every protein encoded by a given genome. This genome-based approach allows for a high-throughput method of structure determination by a combination of experimental and modeling approaches. The principal difference between structural genomics and traditional structural prediction is that structural genomics attempts to determine the structure of every protein encoded by the genome, rather than focusing on one particular protein. With full-genome sequences available, structure prediction can be done more quickly through a combination of experimental and modeling approaches, especially because the availability of large numbers of sequenced genomes and previously solved protein structures allow scientists to model protein structure on the structures of previously solved homologs. Structural genomics involves taking a large number of approaches to structure determination, including experimental methods using genomic sequences or modeling-based approaches based on sequence or structural homology to a protein of known structure or based on chemical and physical principles for a protein with no homology to any known structure. As opposed to traditional structural biology, the determination of a protein structure through a structural genomics effort often (but not always) comes before anything is known regarding the protein function. This raises new challenges in structural bioinformatics, i.e. determining protein function from its 3D structure.

Epigenomics

Epigenomics is the study of the complete set of epigenetic modifications on the genetic material of a cell, known as the epigenome. Epigenetic modifications are reversible modifications on a cell's DNA or histones that affect gene expression without altering the DNA sequence. Two of the most characterized epigenetic modifications are DNA methylation and histone modification. Epigenetic modifications play an important role in gene expression and regulation, and are involved in numerous cellular processes such as in differentiation/development and tumorigenesis. The study of epigenetics on a global level has been made possible only recently through the adaptation of genomic high-throughput assays.

Metagenomics

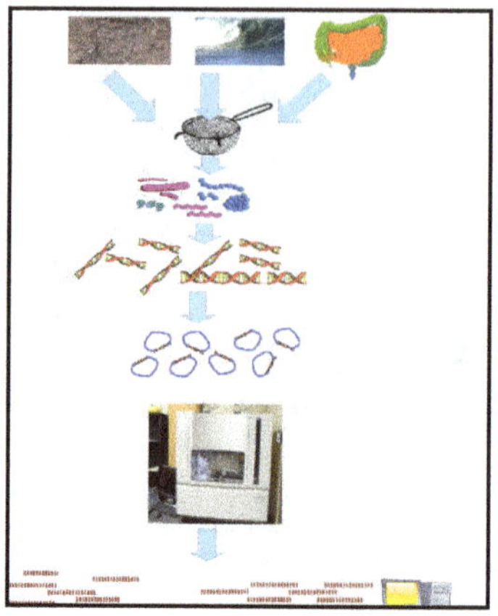

Environmental Shotgun Sequencing (ESS) is a key technique in metagenomics. (A) Sampling from habitat; (B) filtering particles, typically by size; (C) Lysis and DNA extraction; (D) cloning and library construction; (E) sequencing the clones; (F) sequence assembly into contigs and scaffolds.

Metagenomics is the study of *metagenomes*, genetic material recovered directly from environmental samples. The broad field may also be referred to as environmental genomics, ecogenomics or community genomics. While traditional microbiology and microbial genome sequencing rely upon cultivated clonal cultures, early environmental gene sequencing cloned specific genes (often the 16S rRNA gene) to produce a profile of diversity in a natural sample. Such work revealed that the vast majority of microbial biodiversity had been missed by cultivation-based methods. Recent studies use "shotgun" Sanger sequencing or massively parallel pyrosequencing to get largely unbiased samples of all genes from all the members of the sampled communities. Because of its power to reveal the previously hidden diversity of microscopic life, metagenomics offers a powerful lens for viewing the microbial world that has the potential to revolutionize understanding of the entire living world.

Study Systems

Viruses and Bacteriophages

Bacteriophages have played and continue to play a key role in bacterial genetics and molecular biology. Historically, they were used to define gene structure and gene regulation. Also the first genome to be sequenced was a bacteriophage. However, bacteriophage research did not lead the genomics revolution, which is clearly dominated by bacterial genomics. Only very recently has the study of bacteriophage genomes become prominent, thereby enabling researchers to understand the mechanisms underlying phage evolution. Bacteriophage genome sequences can be obtained through direct sequencing of isolated bacteriophages, but can also be derived as part of microbial genomes. Analysis of bacterial genomes has shown that a substantial amount of microbial DNA consists of prophage sequences and prophage-like elements. A detailed database mining of these sequences offers insights into the role of prophages in shaping the bacterial genome.

Cyanobacteria

At present there are 24 cyanobacteria for which a total genome sequence is available. 15 of these cyanobacteria come from the marine environment. These are six *Prochlorococcus* strains, seven marine *Synechococcus* strains, *Trichodesmium erythraeum* IMS101 and *Crocosphaera watsonii* WH8501. Several studies have demonstrated how these sequences could be used very successfully to infer important ecological and physiological characteristics of marine cyanobacteria. However, there are many more genome projects currently in progress, amongst those there are further *Prochlorococcus* and marine *Synechococcus* isolates, *Acaryochloris* and *Prochloron*, the N_2-fixing filamentous cyanobacteria *Nodularia spumigena*, *Lyngbya aestuarii* and *Lyngbya majuscula*, as well as bacteriophages infecting marine cyanobaceria. Thus, the growing body of genome information can also be tapped in a more general way to address global problems by applying a comparative approach. Some new and exciting examples of progress in this field are the identification of genes for regulatory RNAs, insights into the evolutionary origin of photosynthesis, or estimation of the contribution of horizontal gene transfer to the genomes that have been analyzed.

Human Genomics

Applications of Genomics

Genomics has provided applications in many fields, including medicine, biotechnology, anthropology and other social sciences.

Genomic Medicine

Next-generation genomic technologies allow clinicians and biomedical researchers to drastically increase the amount of genomic data collected on large study populations. When combined with new informatics approaches that integrate many kinds of data with genomic data in disease research, this allows researchers to better understand the genetic bases of drug response and disease.

Synthetic Biology and Bioengineering

The growth of genomic knowledge has enabled increasingly sophisticated applications of synthetic biology. In 2010 researchers at the J. Craig Venter Institute announced the creation of a partially synthetic species of bacterium, *Mycoplasma laboratorium*, derived from the genome of *Mycoplasma genitalium*.

Conservation Genomics

Conservationists can use the information gathered by genomic sequencing in order to better evaluate genetic factors key to species conservation, such as the genetic diversity of a population or whether an individual is heterozygous for a recessive inherited genetic disorder. By using genomic data to evaluate the effects of evolutionary processes and to detect patterns in variation throughout a given population, conservationists can formulate plans to aid a given species without as many variables left unknown as those unaddressed by standard genetic approaches.

Personal Genomics

Personal genomics is the branch of genomics concerned with the sequencing and analysis of the genome of an individual. The genotyping stage employs different techniques including single-nucleotide polymorphism (SNP) analysis chips (typically 0.02% of the genome), or partial or full genome sequencing. Once the genotypes are known, the individual's genotype can be compared with the published literature to determine likelihood of trait expression and disease risk.

Automated sequencers have increased the speed and reduced the cost of sequencing making it possible to offer genetic testing to consumers.

Use of Personal Genomics in Predictive and Precision Medicine

Predictive medicine is the use of the information produced by personal genomics techniques when deciding what medical treatments are appropriate for a particular individual. Precision medicine is focused on "a new taxonomy of human disease based on molecular biology".

Examples of the use of predictive and precision medicine include inherited medical genomics, cancer genomics and pharmacogenomics. In pharmacogenomics genetic information can be used to select the most appropriate drug to prescribe to a patient. The drug should be chosen to maximize the probability of obtaining the desired result in the patient and minimize the probability that the patient will experience side effects. Genetic information may allow physicians to tailor therapy to a given patient, in order to increase drug efficacy and minimize side effects. As of Oct 2012 there are 167 examples of drug gene pairs for which this information is currently useful in clinical practice and this number has been growing rapidly.

Disease risk may be calculated based on genetic markers and genome-wide association studies for common medical conditions, which are multifactorial and include environmental components in the assessment. Diseases which are individually rare (less than 200,000 people affected in the USA) are nevertheless collectively common (affecting roughly 8-10% of the US population). Over 2500 of these diseases (including a few more common ones) have predictive genetics of sufficiently high clinical impact that they are recommended as medical genetic tests available for single genes (and in whole genome sequencing) and growing at about 200 new genetic diseases per year.

Cost of Sequencing an Individual's Genome

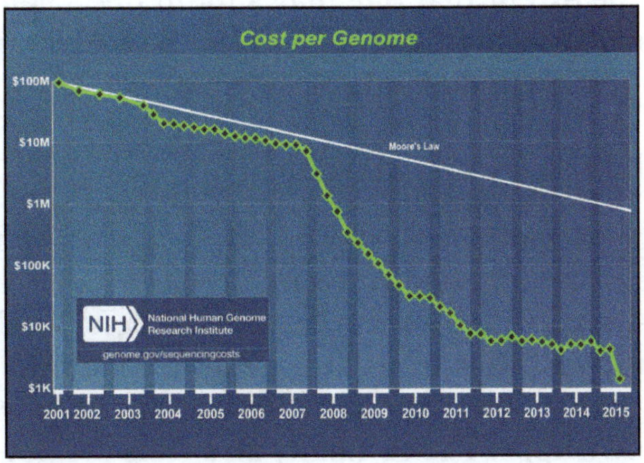

Typical cost of sequencing a human-sized genome, on a logarithmic scale. Note the drastic trend faster than Moore's law beginning in January 2008 as post-Sanger sequencing came online at sequencing centers.

The cost of sequencing a human genome is dropping rapidly, due to the continual development of new, faster, cheaper DNA sequencing technologies such as "next generation DNA sequencing".

The National Human Genome Research Institute, part of the U.S. National Institutes of Health, set a target to be able to sequence a human-sized genome for US$100,000 by 2009 and US$1,000 by 2014.

There are 6 billion base pairs in the diploid human genome. Statistical analysis reveals that a coverage of approximately ten times is required to get coverage of both alleles in 90% human genome from 25 base pair reads with shotgun sequencing. This means a total of 60 billion base pairs that must be sequenced. An Applied Biosystems SOLiD, Illumina or Helicos sequencing machine can sequence 2 to 10 billion base pairs in each $8,000 to $18,000 run. The cost must also take into account personnel costs, data processing costs, legal, communications and other costs. One way to assess this is via commercial offerings. The first such whole diploid genome sequencing (6 billion bp, 3 billion from each parent) was from Knome and their price dropped from $350,000 in 2008 to $99,000 in 2009. This inspects 3000-fold more bases of the genome than SNP chip-based genotyping, identifying both novel and known sequence variants, some relevant to personal health or ancestry. In June 2009, Illumina announced the launch of its own Personal Full Genome Sequencing Service at a depth of 30X for $48,000 per genome. In 2010, they cut the price to $19,500.

In 2009, Complete Genomics of Mountain View announced that it would provide full genome sequencing for $5,000, from June 2009. This will only be available to institutions, not individuals. Prices are expected to drop further over the next few years through economies of scale and increased competition. As of 2014, full exome sequencing was offered by Gentle for less than $2,000, including personal counseling along with the results.

The decreasing cost in general of genomic mapping has permitted genealogical sites to offer it as a service, to the extent that one may submit one's genome to crowd sourced scientific endeavours such as DNA.land at the New York Genome Center, an example both of the economies of scale and of citizen science.

Projects and Services Already Available

- Sequencing.com provides free, unlimited data storage for all genetic data in high-security data centers and software applications to analyze the data based on their patent-pending Real-Time Personalization™ technology. The company invented an open API that translates genetic code into software code so that third-party developers without any training in genetics can integrate genetic code into their own apps. The company also runs the not-for-profit Altruist Endeavor, which is an open data initiative consisting of a free, publicly accessible online repository of anonymous genetic data in order to enable genetic research.
- The Genographic Project is a project of the National Geographic Society and IBM to collect DNA samples to map historical human migration patterns. Launched in 2005, with more than 650,000 public participants as of December 2015, it helped to create the direct-to-consumer (DTC) genetic testing industry.
- The Personal Genome Project (PGP) is a long term, large cohort study based at Harvard Medical School which aims to sequence and publicize the complete genomes and medical records of 100,000 volunteers, in order to enable research into personal genomics and personalized medicine.
- SNPedia is a wiki that collects and shares information about the consequences of DNA variations, and through the associated program Promethease, anyone who has obtained DNA data about themselves (from any company) can get a free, independent report containing risk assessments and related information.
- deCODEme.com charged $1100 to carry out genotyping of approximately 1 million SNPs and provided risk estimates for 47 diseases as well as ancestry analyses. However, sales of genetic scans direct to consumer through deCODEme have now been discontinued.
- Navigenics began offering SNP-based genomic risk assessments as of April 2008. Navigenics is medically focused and emphasizes a clinician's and genetic counselor's role in interpreting results. Affymetrix Genome-Wide Human SNP Array 6.0, which genotypes 900,000 SNPs. Navigenics' service has now been discontinued.
- Pathway Genomics analyzes over 100 genetic markers to identify genetic risk for common health conditions such as melanoma, prostate cancer and rheumatoid arthritis.

- LifeNome provides an analysis of over 6500 genetic variations to assess predisposition likelihood for over 800 complex wellness traits across nutrition and dieting, fitness, skin care, preventive well-being, aging, allergies and sensitivities and other wellness traits.
- UBiome is a bacterial genomics company who offer microbiome sequencing to the public based in San Francisco, California.
- 23andMe founded in 2006 allows people to send their saliva in a testing kit to have their DNA screened for genes associated with certain inherited conditions, such as cystic fibrosis or sickle cell anaemia, and other genetic markers relating to parts of their lives and ancestry. Its saliva-based direct-to-consumer personal genome test was named Invention of the Year by *Time* magazine in 2008. The company is now backed by Google.
- Illumina, Oxford Nanopore Technologies, Sequenom, Pacific Biosciences, Complete Genomics, and 454 Life Sciences are commercializing full genome sequencing but do not provide genetic analysis or counselling.
- DNALifeStyleCoach is a service by Titanovo that offers SNPs genotyping and results interpretation in areas of nutrigenetics, sports genetics, anti-aging and psychogenetics.
- Gene by Gene provides Whole Genome Sequencing from $6995 to $7595, and other variants.
- Life Technologies.
- Mapmygenome offers several SNP and NGS panels for personal genomics.
- Full Genomes Corporation offers whole genome sequencing and y chromosome sequencing for personal genomics.

Ethical Issues

Genetic discrimination is discriminating on the basis of information obtained from an individual's genome. Genetic non-discrimination laws have been enacted in some US states and at the federal level, by the Genetic Information Nondiscrimination Act (GINA). The GINA legislation prevents discrimination by health insurers and employers, but does not apply to life insurance or long-term care insurance. Given the ethical concerns about presymptomatic genetic testing of minors, it is likely that personal genomics will first be applied to adults who can provide consent to undergo such testing, although genome sequencing is already proving valuable for children if any symptoms are present.

Patients will need to be educated on interpreting their results and what they should be rationally taking from the experience. It is not only the average person who needs to be educated in the dimensions of their own genomic sequence but also professionals, including physicians and science journalists, who must be provided with the knowledge required to inform and educate their patients and the public Examples of such efforts include the Personal Genetics Education Project (pgEd) and the Smithsonian collaboration with NHGRI.

Other Issues

Full sequencing of the genome can identify polymorphisms that are so rare that no conclusions may be drawn about their impact, creating uncertainty in the analysis of individual genomes, particularly in the context of clinical care. Czech medical geneticist Eva Machácková writes: “In some cases it is difficult to distinguish if the detected sequence variant is a causal mutation or a neutral (polymorphic) variation without any effect on phenotype. The interpretation of rare sequence variants of unknown significance detected in disease-causing genes becomes an increasingly important problem.”

There is a heavy debate as to how relevant the results of personal genome kits are and whether or not the ramifications of knowing one’s predisposition to a disease is worth the potential psychological stress. There are also three potential problems associated with the validity of personal genome kits. The first issue is the test’s validity. Handling errors of the sample increases the likelihood for errors which could affect the test results and interpretation. The second affects the clinical validity, which could affect the test’s ability to detect or predict associated disorders. The third problem is the clinical utility of personal genome kits and associated risks, and the benefits of introducing them into clinical practices.

Doctors are currently conducting tests for which some are not correctly trained to interpret the results. Many are unaware of how SNPs respond to one another. This results in presenting the client with potentially misleading and worrisome results which could strain the already overloaded health care system. This may antagonize the individual to make uneducated decisions such as unhealthy lifestyle choices and family planning modifications. Moreover, negative results which may potentially be inaccurate, theoretically decrease the quality of life and mental health of the individual (such as increased depression and extensive anxiety).

There is also controversy regarding the concerns with companies testing individual DNA. There are issues such as “leaking” information, the right to privacy and what responsibility the company has to ensure this does not happen. Regulation rules are not clearly laid out. What is still not determined is who legally owns the genome information: the company or the individual whose genome has been read. There have been published examples of personal genome information being exploited. Additional privacy concerns, related to, e.g., genetic discrimination, loss of anonymity, and psychological impacts, have been increasingly pointed out by the academic community as well as government agencies.

Conversely, sequencing one's genome would allow for more personalized medical treatments using pharmacogenomics; the use of genetic information to select appropriate drugs. Treatments can be catered to the individual and the certain genetic predispositions they may have (such as personalized chemotherapy).

Popular Culture

The 1997 science fiction film GATTACA presents a near-future society where personal genomics is readily available to anyone, and explores its societal impact.

Oncogenomics

Oncogenomics is a sub-field of genomics that characterizes cancer-associated genes. It focuses on genomic, epigenomic and transcript alterations in cancer.

Cancer is a genetic disease caused by accumulation of DNA mutations and epigenetic alterations leading to unrestrained cell proliferation and neoplasm formation. The goal of oncogenomics is to identify new oncogenes or tumor suppressor genes that may provide new insights into cancer diagnosis, predicting clinical outcome of cancers and new targets for cancer therapies. The success of targeted cancer therapies such as Gleevec, Herceptin and Avastin raised the hope for oncogenomics to elucidate new targets for cancer treatment.

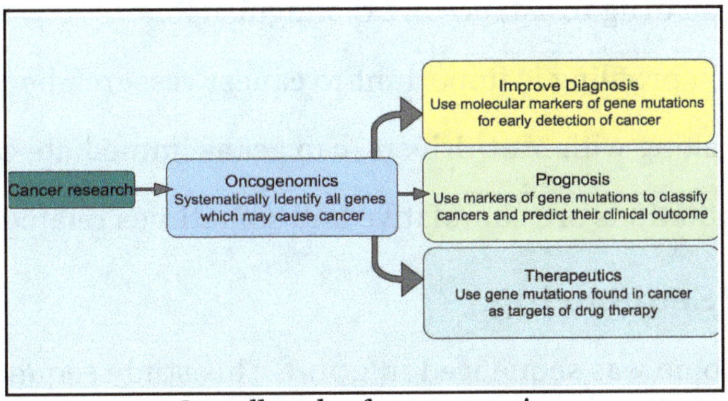

Overall goals of oncogenomics

Besides understanding the underlying genetic mechanisms that initiate or drive cancer progression, oncogenomics targets personalized cancer treatment. Cancer develops due to DNA mutations and epigenetic alterations that accumulate randomly. Identifying and targeting the mutations in an individual patient may lead to increased treatment efficacy.

The completion of the Human Genome Project facilitated the field of oncogenomics and increased the abilities of researchers to find oncogenes. Sequencing technologies and global methylation profiling techniques have been applied to the study of oncogenomics.

History

The genomics era began in the 1990s, with the generation of DNA sequences of many organisms. In the 21st century, the completion of the Human Genome Project enabled the study of functional genomics and examining tumor genomes. Cancer is a main focus.

The epigenomics era largely began more recently, about 2000. One major source of epigenetic change is altered methylation of CpG islands at the promoter region of genes. A number of recently devised methods can assess the DNA methylation status in cancers versus normal tissues. Some methods assess methylation of CpGs located in different classes of loci, including CpG islands, shores, and shelves as well as promoters, gene bodies, and intergenic regions. Cancer is also a major focus of epigenetic studies.

Access to whole cancer genome sequencing is important to cancer (or cancer genome) research because:

- Mutations are the immediate cause of cancer and define the tumor phenotype.
- Access to cancerous and normal tissue samples from the same patient and the fact that most cancer mutations represent somatic events, allow the identification of cancer-specific mutations.
- Cancer mutations are cumulative and sometimes are related to disease stage. Metastasis and drug resistance are distinguishable.

Access to methylation profiling is important to cancer research because:

- Epi-drivers, along with Mut-drivers, can act as immediate causes of cancers.
- Cancer epimutations are cumulative and sometimes related to disease stage.

Whole Genome Sequencing

The first cancer genome was sequenced in 2008. This study sequenced a typical acute myeloid leukaemia (AML) genome and its normal counterpart genome obtained from the same patient. The comparison revealed ten mutated genes. Two were already thought to contribute to tumor progression: an internal tandem duplication of the FLT3 receptor tyrosine kinase gene, which activates kinase signaling and is associated with a poor prognosis and a four base insertion in exon 12 of the NPM1 gene (NPMc). These mutations are found in 25-30% of AML tumors and are thought to contribute to disease progression rather than to cause it directly.

The remaining 8 were new mutations and all were single base changes: Four were in families that are strongly associated with cancer pathogenesis (PTPRT, CDH24, PCLKC and SLC15A1). The other four had no previous association with cancer pathogenesis. They did have potential functions in metabolic pathways that suggested mechanisms by which they could act to promote cancer (KNDC1, GPR124, EB12, GRINC1B).

These genes are involved in pathways known to contribute to cancer pathogenesis, but before this study most would not have been candidates for targeted gene therapy. This analysis validated the approach of whole cancer genome sequencing in identifying somatic mutations and the importance of parallel sequencing of normal and tumor cell genomes.

In 2011, the genome of an exceptional bladder cancer patient whose tumor had been eliminated by the drug everolimus was sequenced, revealing mutations in two genes, *TSC1* and *NF2*. The mutations disregulated mTOR, the protein inhibited by everolimus, allowing it to reproduce without limit. As a result, in 2015, the Exceptional Responders Initiative was created at the National Cancer Institute. The initiative allows such exceptional patients (who have responded positively for at least six months to a cancer drug that usually fails) to have their genomes sequenced to identify the relevant mutations. Once identified, other patients could be screened for those mutations and then be given the drug. In 2016 To that end, a nationwide cancer drug trial began in 2015, involving up to twenty-four hundred centers. Patients with appropriate mutations are matched with one of more than forty drugs.

In 2014 the Center for Molecular Oncology rolled out the MSK-IMPACT test, a screening tool that looks for mutations in 341 cancer-associated genes. By 2015 more than five thousand patients had been screened. Patients with appropriate mutations are eligible to enroll in clinical trials that provide targeted therapy.

Technologies

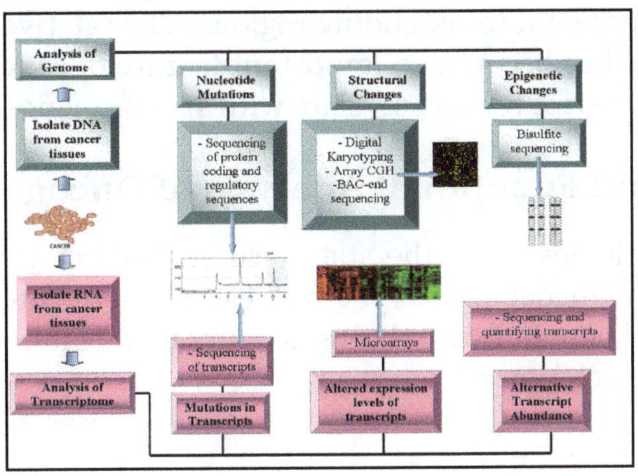

Current technologies being used in Oncogenomics.

Genomics technologies include:

Genome Sequencing

- *DNA sequencing*: Pyrosequencing-based sequencers offer a relatively low-cost method to generate sequence data.
- *Array Comparative Genome Hybridization*: This technique measures the DNA copy number differences between normal and cancer genomes. It uses the fluorescence intensity from fluorescent-labeled samples, which are hybridized to known probes on a microarray.
- *Representational oligonucleotide microarray analysis*: Detects copy number variation using amplified restriction-digested genomic fragments that are hybridized to human oligonucleotides, achieving a resolution between 30 and 35 kbit/s.
- *Digital Karyotyping*: Detects copy number variation using genomics tags obtained via restriction enzyme digests. These tags are then linked to into ditags, concatenated, cloned, sequenced and mapped back to the reference genome to evaluate tag density.
- *Bacterial Artificial Chromosome (BAC)-end sequencing (end-sequence profiling)*: Identifies chromosomal breakpoints by generating a BAC library from a cancer genome and sequencing their ends. The BAC clones that contain chromosome aberrations have end sequences that do not map to a similar region of the reference genome, thus identifying a chromosomal breakpoint.

Transcriptomes

- *Microarrays*: Assess transcript abundance. Useful in classification, prognosis, raise the possibility of differential treatment approaches and aid identification of mutations in the proteins' coding regions. The relative abundance of alternative transcripts has become an important feature of cancer research. Particular alternative transcript forms correlate with specific cancer types.

Bioinformatics and Functional Analysis of Oncogenes

Bioinformatics technologies allow the statistical analysis of genomic data. The functional characteristics of oncogenes has yet to be established. Potential functions include their transformational capabilities relating to tumour formation and specific roles at each stage of cancer development.

Operomics

Operomics aims to integrate genomics, transcriptomics and proteomics to understand the molecular mechanisms that underlie the cancer development.

Comparative Oncogenomics

Comparative oncogenomics uses cross-species comparisons to identify oncogenes. This research involves studying cancer genomes, transcriptomes and proteomes in model organisms such as mice, identifying potential oncogenes and referring back to human cancer samples to see whether homologues of these oncogenes are important in causing human cancers. Genetic alterations in mouse models are similar to those found in human cancers. These models are generated by methods including retroviral insertion mutagenesis or graft transplantation of cancerous cells.

Source of Carcinogenic Driver Mutations

As pointed out by Gao et al., the stability and integrity of the human genome are maintained by the DNA damage repair (DDR) system. Un-repaired DNA damage is a major cause of mutations that drive carcinogenesis. If DNA repair is deficient, DNA damage tends to accumulate. Such excess DNA damage can increase mutational errors during DNA replication due to error-prone translesion synthesis. Excess DNA damage can also increase epigenetic alterations due to errors during DNA repair. Such mutations and epigenetic alterations can give rise to cancer.

DDR genes are often repressed in human cancer by epigenetic mechanisms. Such repression may involve DNA methylation of promoter regions or repression of DDR genes by a microRNA. Epigenetic repression of DDR genes occurs more frequently than gene mutation in many types of cancer. Thus, epigenetic repression often plays a more important role than mutation in reducing expression of DDR genes. This reduced expression of DDR genes is likely an important driver of carcinogenesis.

Synthetic Lethality

Synthetic lethality arises when a combination of deficiencies in the expression of two or more genes leads to cell death, whereas a deficiency in only one of these genes does not. The deficiencies can arise through mutations, epigenetic alterations or inhibitors of one of the genes.

Some oncogenes are essential for survival of all cells (not only cancer cells). Thus, drugs that knock out these oncogenes (and thereby kill cancer cells) may also damage normal cells, inducing significant illness. However, other genes may be essential to cancer cells but not to healthy cells.

Treatments based on the principle of synthetic lethality have prolonged the survival of cancer patients, and show promise for future advances in reversal of carcinogenesis. A major type of synthetic lethality operates on the DNA repair defect that often initiates a cancer, and is still present in the tumor cells. Some examples are given here.

BRCA1 or BRCA2 expression is deficient in a majority of high-grade breast and ovarian cancers, usually due to epigenetic methylation of its promoter or epigenetic repression by an over-expressed microRNA. BRCA1 and BRCA2 are important components of the major pathway for homologous recombinational repair of double-strand breaks. If one or the other is deficient, it increases the risk of cancer, especially breast or ovarian cancer. A back-up DNA repair pathway, for some of the damages usually repaired by BRCA1 and BRCA2, depends on PARP1. Thus, many ovarian cancers respond to an FDA-approved treatment with a PARP inhibitor, causing synthetic lethality to cancer cells deficient in BRCA1 or BRCA2. This treatment is also being evaluated for breast cancer and numerous other cancers in Phase III clinical trials in 2016.

There are two pathways for homologous recombinational repair of double-strand breaks. The major pathway depends on BRCA1, PALB2 and BRCA2 while an alternative pathway depends on RAD52. Pre-clinical studies, involving epigenetically reduced or mutated BRCA-deficient cells (in culture or injected into mice), show that inhibition of RAD52 is synthetically lethal with BRCA-deficiency.

Mutations in genes employed in DNA mismatch repair (MMR) cause a high mutation rate. In tumors, such frequent subsequent mutations often generate "non-self" immunogenic antigens. A human Phase II clinical trial, with 41 patients, evaluated one synthetic lethal approach for tumors with or without MMR defects. The product of gene *PD-1* ordinarily represses cytotoxic immune responses. Inhibition of this gene allows a greater immune response. When cancer patients with a defect in MMR in their tumors were exposed to an inhibitor of PD-1, 67% - 78% of patients experienced immune-related progression-free survival. In contrast, for patients without defective MMR, addition of PD-1 inhibitor generated only 11% of patients with immune-related progression-free survival. Thus inhibition of PD-1 is primarily synthetically lethal with MMR defects.

ARID1A, a chromatin modifier, is required for non-homologous end joining, a major pathway that repairs double-strand breaks in DNA, and also has transcription regulatory roles. ARID1A mutations are one of the 12 most common carcinogenic mutations. Mutation or epigenetically decreased expression of ARID1A has been found in 17 types of cancer. Pre-clinical studies in cells and in mice show that synthetic lethality for ARID1A deficiency occurs by either inhibition of the methyltransferase activity of EZH2, or with addition of the kinase inhibitor dasatinib.

Another approach is to individually knock out each gene in a genome and observe the effect on normal and cancerous cells. If the knockout of an otherwise nonessential gene has little or no effect on healthy cells, but is lethal to cancerous cells containing a mutated oncogene, then the system-wide suppression of the suppressed gene can destroy cancerous cells while leaving healthy ones relatively undamaged. The technique was used to identify PARP-1 inhibitors to treat BRCA1/BRCA2-associated cancers. In this case, the combined presence of PARP-1 inhibition and of the cancer-associated mutations in BRCA genes is lethal only to the cancerous cells.

Databases for Cancer Research

The Cancer Genome Project is an initiative to map out all somatic mutations in cancer. The project systematically sequences the exons and flanking splice junctions of the genomes of primary tumors and cancerous cell lines. COSMIC software displays the data generated from these experiments. As of February 2008, the CGP had identified 4,746 genes and 2,985 mutations in 1,848 tumours.

The Cancer Genome Anatomy Project includes information of research on cancer genomes, transcriptomes and proteomes.

Progenetix is an oncogenomic reference database, presenting cytogenetic and molecular-cytogenetic tumor data.

Oncomine has compiled data from cancer transcriptome profiles.

The integrative oncogenomics database IntOGen and the Gitools datasets integrate multidimensional human oncogenomic data classified by tumor type. The first version of IntOGen focused on the role of deregulated gene expression and CNV in cancer. A later version emphasized mutational cancer driver genes across 28 tumor types,. All releases of IntOGen data are made available at the IntOGen database.

The International Cancer Genome Consortium is the biggest project to collect human cancer genome data. The data is accessible through the ICGC website. The BioExpress® Oncology Suite contains gene expression data from primary, metastatic and benign tumor samples and normal samples, including matched adjacent controls. The suite includes hematological malignancy samples for many well-known cancers.

Specific databases for model animals include the Retrovirus Tagged Cancer Gene Database (RTCGD) that compiled research on retroviral and transposon insertional mutagenesis in mouse tumors.

Gene Families

Mutational analysis of entire gene families revealed that genes of the same family have similar functions, as predicted by similar coding sequences and protein domains. Two such classes are the kinase family, involved in adding phosphate groups to proteins and the phosphatase family, involved with removing phosphate groups from proteins. These families were first examined because of their apparent role in transducing cellular signals of cell growth or death. In particular, more than 50% of colorectal cancers carry a mutation in a kinase or phosphatase gene. Phosphatidylinositold 3-kinases (PIK3CA) gene encodes for lipid kinases that commonly contain mutations in colorectal, breast, gastric, lung and various other cancers. Drug therapies can inhibit PIK3CA. Another example is the BRAF gene, one of the first to be implicated in melanomas. BRAF encodes a serine/threonine kinase that is involved in the RAS-RAF-MAPK growth signaling pathway. Mutations in BRAF cause constitutive phosphorylation and activity in

59% of melanomas. Before BRAF, the genetic mechanism of melanoma development was unknown and therefore prognosis for patients was poor.

Mitochondrial DNA

Mitochondrial DNA (mtDNA) mutations are linked the formation of tumors. Four types of mtDNA mutations have been identified:

Point Mutations

Point mutations have been observed in the coding and non-coding region of the mtDNA contained in cancer cells. In individuals with bladder, head/neck and lung cancers, the point mutations within the coding region show signs of resembling each other. This suggests that when a healthy cell transforms into a tumor cell (a neoplastic transformation) the mitochondria seem to become homogenous. Abundant point mutations located within the non-coding region, D-loop, of the cancerous mitochondria suggest that mutations within this region might be an important characteristic in some cancers.

Deletions

This type of mutation is sporadically detected due to its small size (< 1kb). The appearance of certain specific mtDNA mutations (264-bp deletion and 66-bp deletion in the complex 1 subunit gene ND1) in multiple types of cancer provide some evidence that small mtDNA deletions might appear at the beginning of tumorigenesis. It also suggests that the amount of mitochondria containing these deletions increases as the tumor progresses. An exception is a relatively large deletion that appears in many cancers (known as the "common deletion"), but more mtDNA large scale deletions have been found in normal cells compared to tumor cells. This may be due to a seemingly adaptive process of tumor cells to eliminate any mitochondria that contain these large scale deletions (the "common deletion" is > 4kb).

Insertions

Two small mtDNA insertions of ~260 and ~520 bp can be present in breast cancer, gastric cancer, hepatocellular carcinoma (HCC) and colon cancer and in normal cells. No correlation between these insertions and cancer are established.

Copy Number Mutations

The characterization of mtDNA via real-time polymerase chain reaction assays shows the presence of quantitative alteration of mtDNA copy number in many cancers. Increase in copy number is expected to occur because of oxidative stress. On the other hand, decrease is thought to be caused by somatic point mutations in the replication origin site of the H-strand and/or the D310 homopolymeric c-stretch in the D-loop region, mutations in the p53 (tumor suppressor gene) mediated pathway and/or inefficient enzyme activity due to POLG mutations. Any increase/decrease in copy number then remains constant within tumor cells. The fact that the amount of mtDNA is constant in tumor cells suggests that the amount of mtDNA is controlled by a much more complicated system in tumor cells, rather than simply altered as a consequence of abnormal cell proliferation. The role of mtDNA content in human cancers apparently varies for particular tumor types or sites.

Mutations in mitochondrial DNA in various cancers

Cancer Type	**Location of Point mutations**				**Nucleotide Position of Deletions**	**Increase of mtDNA copy #**	**Decrease of mtDNA copy #**
	D-Loop	**mR-NAs**	**tR-NAs**	**rR-NAs**			
Bladder	X	X		X	15,642-15,662		
Breast	X	X	X	X	8470-13,447 and 8482-13459		X
Head and neck	X	X	X	X	8470-13,447 and 8482-13459	X	
Oral	X	X			8470-13,447 and 8482-13459		
Hepatocellular carcinoma (HCC)	X	X	X	X	306-556 and 3894-3960		X
Esophageal	X	X		X	8470-13,447 and 8482-13459	X	
Gastric	X	X	X		298-348		X
Prostate	X			X	8470-13,447 and 8482-13459	X	

57.7% (500/867) contained somatic point putations and of the 1172 mutations surveyed 37.8% (443/1127) were located in the D-loop control region, 13.1% (154/1172) were located in the tRNA or rRNA genes and 49.1% (575/1127) were found in the mRNA genes needed for producing complexes required for mitochondrial respiration.

Diagnostic Applications

Some anticancer drugs target mtDNA and have shown positive results in killing tumor cells. Research has used mitochondrial mutations as biomarkers for cancer cell therapy. It is easier to target mutation within mitochondrial DNA versus nuclear DNA because the mitochondrial genome is much smaller and easier to screen for specific mutations. MtDNA content alterations found in blood samples might be able to serve as a screening marker for predicting future cancer susceptibility as well as tracking malignant tumor progression. Along with these potential helpful characteristics of mtDNA, it is not under the control of the cell cycle and is important for maintaining ATP generation and mitochondrial homeostasis. These characteristics make targeting mtDNA a practical therapeutic strategy.

Cancer Biomarkers

Several biomarkers can be useful in cancer staging, prognosis and treatment. They can range from single-nucleotide polymorphisms (SNPs), chromosomal aberrations, changes in DNA copy number, microsatellite instability, promoter region methylation, or even high or low protein levels.

Comparative Genomics

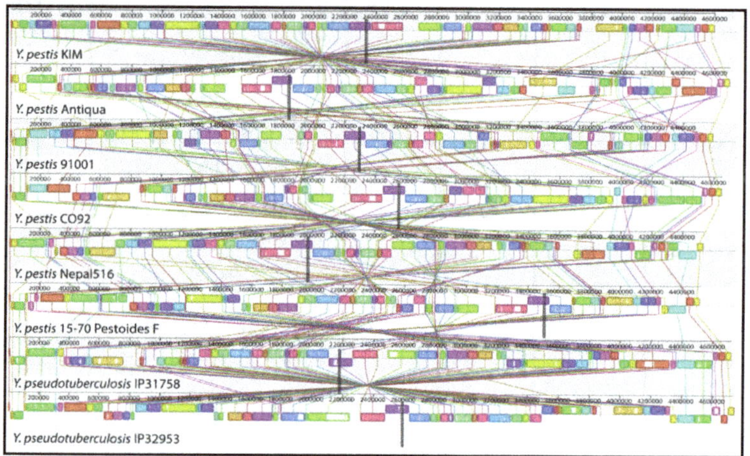

Whole genome alignment is a typical method in comparative genomics. This alignment of eight *Yersinia* bacteria genomes reveals 78 locally collinear blocks conserved among all eight taxa. Each chromosome has been laid out horizontally and homologous blocks in each genome are shown as identically colored regions linked across genomes. Regions that are inverted relative to *Y. pestis* KIM are shifted below a genome's center axis.

Comparative genomics is a field of biological research in which the genomic features of different organisms are compared. The genomic features may include the DNA sequence, genes, gene order, regulatory sequences, and other genomic structural landmarks. In this branch of genomics, whole or large parts of genomes resulting from genome projects are compared to study basic biological similarities and differences as well as evolutionary relationships between organisms. The major principle of comparative genomics is that common features of two organisms will often be encoded within the DNA that is evolutionarily conserved between them. Therefore, comparative genomic approaches start with making some form of alignment of genome sequences and looking for orthologous sequences (sequences that share a common ancestry) in the aligned genomes and checking to what extent those sequences are conserved. Based on these, genome and molecular evolution are inferred and this may in turn be put in the context of, for example, phenotypic evolution or population genetics.

Virtually started as soon as the whole genomes of two organisms became available (that is, the genomes of the bacteria *Haemophilus influenzae* and *Mycoplasma genitalium*) in 1995, comparative genomics is now a standard component of the analysis of every new genome sequence. With the explosion in the number of genome project due to the advancements in DNA sequencing technologies, particularly the next-generation sequencing methods in late 2000s, this field has become more sophisticated, making it possible to deal with many genomes in a single study. Comparative genomics has revealed high levels of similarity between closely related organisms, such as humans and chimpanzees, and, more surprisingly, similarity between seemingly distantly related organisms, such as humans and the yeast *Saccharomyces cerevisiae*. It has also showed the extreme diversity of the gene composition in different evolutionary lineages.

History

Comparative genomics has a root in the comparison of virus genomes in the early 1980s. For example, small RNA viruses infecting animals (picornaviruses) and those infecting plants (cowpea mosaic virus) were compared and turned out to share significant sequence similarity and, in part, the order of their genes. In 1986, the first comparative genomic study at a larger scale was published, comparing the genomes of varicella-zoster virus and Epstein-Barr virus that contained more than 100 genes each.

The first complete genome sequence of a cellular organism, that of *Haemophilus influenzae* Rd, was published in 1995. The second genome sequencing paper was of the small parasitic bacterium *Mycoplasma genitalium* published in the same year. Starting from this paper, reports on new genomes inevitably became comparative-genomic studies.

The first high-resolution whole genome comparison system was developed in 1998 by Art Delcher, Simon Kasif and Steven Salzberg and applied to the comparison of entire highly related microbial organisms with their collaborators at the Institute for Genomic Research (TIGR). The system is called MUMMER and was described in a publication in Nucleic Acids Research in 1999. The system helps researchers to identify large rearrangements, single base mutations, reversals, tandem repeat expansions and other polymorphisms. In bacteria, MUMMER enables the identification of polymorphisms that are responsible for virulence, pathogenicity, and anti-biotic resistance. The system was also applied to the Minimal Organism Project at TIGR and subsequently to many other comparative genomics projects.

Saccharomyces cerevisiae, the baker's yeast, was the first eukaryote to have its complete genome sequence published in 1996. After the publication of the roundworm *Caenorhabditis elegans* genome in 1998 and together with the fruit fly *Drosophila melanogaster* genome in 2000, Gerald M. Rubin and his team published a paper titled "Comparative Genomics of the Eukaryotes", in which they compared the genomes of the eukaryotes *D. melanogaster*, *C. elegans*, and *S. cerevisiae*, as well as the prokaryote *H. influenzae*. At the same time, Bonnie Berger, Eric Lander, and their team published a paper on whole-genome comparison of human and mouse.

With the publication of the large genomes of vertebrates in the 2000s, including human, the Japanese pufferfish *Takifugu rubripes*, and mouse, precomputed results of large genome comparisons have been released for downloading or for visualization in a genome browser. Instead of undertaking their own analyses, most biologists can access these large cross-species comparisons and avoid the impracticality caused by the size of the genomes.

Next-generation sequencing methods, which were first introduced in 2007, have produced an enormous amount of genomic data and have allowed researchers to generate multiple (prokaryotic) draft genome sequences at once. These methods can also quickly uncover single-nucleotide polymorphisms, insertions and deletions by mapping unassembled reads against a well annotated reference genome, and thus provide a list of possible gene differences that may be the basis for any functional variation among strains.

Evolutionary Principles

One character of biology is evolution, evolutionary theory is also the theoretical foundation of comparative genomics, and at the same time the results of comparative genomics unprecedentedly enriched and developed the theory of evolution. When two or more of the genome sequence are compared, you can get the evolutionary relationships of the sequences in a phylogenetic tree. Based on a variety of biological genome data and the study of vertical and horizontal evolution processes, one can understand vital parts of the gene structure and its regulatory function. But in a genome about 1.5% ~ 14.5% of the genes are related to the "lateral migration phenomenon", namely the transfer of a gene between populations which can exist at the same time. Thus, the differences in sequences have nothing to do with evolution.

Similarity of related genomes is the basis of comparative genomics. If two creatures have a recent common ancestor, the differences between the two species genomes are evolved from the ancestors' genome. The closer the relationship between two organisms, the higher the similiarities between their genomes. If there is close relationship between them, then their genome will display a linear behaviour (synteny), namely some or all of the genetic sequences are conserved. Thus, the genome sequences can be used to identify gene function, by analyzing their homology (sequence similarity) to genes of known function.

Orthologous sequences are related sequences in different species: a gene exists in the original species, the species divided into two species, so genes in new species are orthologous to the sequence in the original species. Paralogous sequences are separated by gene cloning (gene duplication): if a particular gene in the genome is copied, then the copy of the two sequences is paralogous to the original gene. A pair of orthologous sequences is called orthologous pairs (orthologs), a pair of paralogous sequence is called collateral pairs (paralogs). Orthologous pairs usually have the same or similar function, which is not necessarily the case for collateral pairs. In collateral pairs, the sequences tend to evolve into having different functions.

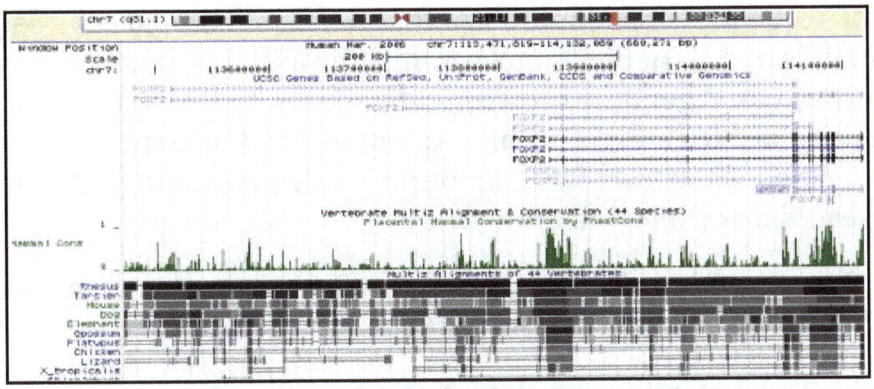

Human FOXP2 gene and evolutionary conservation is shown in and multiple alignment (at bottom of figure) Note that conservation tends to cluster around coding regions (exons).

Comparative genomics exploits both similarities and differences in the proteins, RNA, and regulatory regions of different organisms to infer how selection has acted upon these elements. Those elements that are responsible for similarities between different species should be conserved through time (stabilizing selection), while those elements responsible for differences among species should be divergent (positive selection). Finally, those elements that are unimportant to the evolutionary success of the organism will be unconserved (selection is neutral).

One of the important goals of the field is the identification of the mechanisms of eukaryotic genome evolution. It is however often complicated by the multiplicity of events that have taken place throughout the history of individual lineages, leaving only distorted and superimposed traces in the genome of each living organism. For this reason comparative genomics studies of small model organisms (for example the model Caenorhabditis elegans and closely related Caenorhabditis briggsae) are of great importance to advance our understanding of general mechanisms of evolution.

Methods

Computational approaches to genome comparison have recently become a common research topic in computer science. A public collection of case studies and demonstrations is growing, ranging from whole genome comparisons to gene expression analysis. This has increased the introduction of different ideas, including concepts from systems and control, information theory, strings analysis and data mining. It is anticipated that computational approaches will become and remain a standard topic for research and teaching, while multiple courses will begin training students to be fluent in both topics.

Tools

Computational tools for analyzing sequences and complete genomes are developed quickly due to the availability of large amount of genomic data. At the same time, comparative analysis tools are progressed and improved. In the challenges about these analyses, it is very important to visualize the comparative results.

Visualization of sequence conservation is a tough task of comparative sequence analysis. As we know, it is highly inefficient to examine the alignment of long genomic regions manually. Internet-based genome browsers provide many useful tools for investigating genomic sequences due to integrating all sequence-based biological information on genomic regions. When we extract large amount of relevant biological data, they can be very easy to use and less time-consuming:

- UCSC Browser: This site contains the reference sequence and working draft assemblies for a large collection of genomes.
- Ensembl: The Ensembl project produces genome databases for vertebrates and other eukaryotic species, and makes this information freely available online.
- MapView: The Map Viewer provides a wide variety of genome mapping and sequencing data.
- VISTA is a comprehensive suite of programs and databases for comparative analysis of genomic sequences. It was built to visualize the results of comparative analysis based on DNA alignments. The presentation of comparative data generated by VISTA can easily suit both small and large scale of data.

An advantage of using online tools is that these websites are being developed and updated constantly. There are many new settings and content can be used online to improve efficiency.

Applications

Agriculture

Agriculture is a field that reaps the benefits of comparative genomics. Identifying the loci of advantageous genes is a key step in breeding crops that are optimized for greater yield, cost-efficiency, quality, and disease resistance. For example, one genome wide association study conducted on 517 rice landraces revealed 80 loci associated with several categories of agronomic performance, such as grain weight, amylose content, and drought tolerance. Many of the loci were previously uncharacterized. Not only is this methodology powerful, it is also quick. Previous methods of identifying loci associated with agronomic performance required several generations of carefully monitored breeding of parent strains, a time consuming effort that is unnecessary for comparative genomic studies.

Medicine

The medical field also benefits from the study of comparative genomics. Vaccinology in particular has experienced useful advances in technology due to genomic approaches to problems. In an approach known as reverse vaccinology, researchers can discover candidate antigens for vaccine development by analyzing the genome of a pathogen or a family of pathogens. Applying a comparative genomics approach by analyzing the genomes of several related pathogens can lead to the development of vaccines that are multiprotective. A team of researchers employed such an approach to create a universal vaccine for Group B Streptococcus, a group of bacteria responsible for severe neonatal infection. Comparative genomics can also be used to generate specificity for vaccines against pathogens that are closely related to commensal microorganisms. For example, researchers used comparative genomic analysis of commensal and pathogenic strains of E. coli to identify pathogen specific genes as a basis for finding antigens that result in immune response against pathogenic strains but not commensal ones.

Research

Comparative genomics also opens up new avenues in other areas of research. As DNA sequencing technology has become more accessible, the number of sequenced genomes has grown. With the increasing reservoir of available genomic data, the potency of comparative genomic inference has grown as well. A notable case of this increased potency is found in recent primate research. Comparative genomic methods have allowed researchers to gather information about genetic variation, differential gene expression, and evolutionary dynamics in primates that were indiscernible using previous data and methods. The Great Ape Genome Project used comparative genomic methods to investigate genetic variation with reference to the six great ape species, finding healthy levels of variation in their gene pool despite shrinking population size. Another study showed that patterns of DNA methylation, which are a known regulation mechanism for gene expression, differ in the prefrontal cortex of humans versus chimps, and implicated this difference in the evolutionary divergence of the two species.

Genome Project

Genome projects are scientific endeavours that ultimately aim to determine the complete genome sequence of an organism (be it an animal, a plant, a fungus, a bacterium, an archaean, a protist or a virus) and to annotate protein-coding genes and other important genome-encoded features. The genome sequence of an organism includes the collective DNA sequences of each chromosome in the organism. For a bacterium containing a single chromosome, a genome project will aim to map the sequence of that chromosome. For the human species, whose genome includes 22 pairs of autosomes and 2 sex chromosomes, a complete genome sequence will involve 46 separate chromosome sequences.

When printed, the human genome sequence fills around 100 huge books of close print

The Human Genome Project was a landmark genome project that is already having a major impact on research across the life sciences, with potential for spurring numerous medical and commercial developments.

Genome Assembly

Genome assembly refers to the process of taking a large number of short DNA sequences and putting them back together to create a representation of the original chromosomes from which the DNA originated. In a shotgun sequencing project, all the DNA from a source (usually a single organism, anything from a bacterium to a mammal) is first fractured into millions of small pieces. These pieces are then "read" by automated sequencing machines, which can read up to 1000 nucleotides or bases at a time. (The four bases are adenine, guanine, cytosine, and thymine, represented as AGCT.) A genome assembly algorithm works by taking all the pieces and aligning them to one another, and detecting all places where two of the short sequences, or *reads*, overlap. These overlapping reads can be merged, and the process continues.

Genome assembly is a very difficult computational problem, made more difficult because many genomes contain large numbers of identical sequences, known as repeats. These repeats can be thousands of nucleotides long, and some occur in thousands of different locations, especially in the large genomes of plants and animals.

The resulting (draft) genome sequence is produced by combining the information sequenced contigs and then employing linking information to create scaffolds. Scaffolds are positioned along the physical map of the chromosomes creating a "golden path".

Assembly Software

Originally, most large-scale DNA sequencing centers developed their own software for assembling the sequences that they produced. However, this has changed as the software has grown more complex and as the number of sequencing centers has increased. An example of such assembler *Short Oligonucleotide Analysis Package* developed by

BGI for de novo assembly of human-sized genomes, alignment, SNP detection, resequencing, indel finding, and structural variation analysis.

Genome Annotation

Since the 1980s, molecular biology and bioinformatics have created the need for DNA annotation. DNA annotation or genome annotation is the process of identifying attaching biological information to sequences, and particularly in identifying the locations of genes and determining what those genes do.

When is a Genome Project Finished?

When sequencing a genome, there are usually regions that are difficult to sequence (often regions with highly repetitive DNA). Thus, 'completed' genome sequences are rarely ever complete, and terms such as 'working draft' or 'essentially complete' have been used to more accurately describe the status of such genome projects. Even when every base pair of a genome sequence has been determined, there are still likely to be errors present because DNA sequencing is not a completely accurate process. It could also be argued that a complete genome project should include the sequences of mitochondria and (for plants) chloroplasts as these organelles have their own genomes.

It is often reported that the goal of sequencing a genome is to obtain information about the complete set of genes in that particular genome sequence. The proportion of a genome that encodes for genes may be very small (particularly in eukaryotes such as humans, where coding DNA may only account for a few percent of the entire sequence). However, it is not always possible (or desirable) to only sequence the coding regions separately. Also, as scientists understand more about the role of this noncoding DNA (often referred to as junk DNA), it will become more important to have a complete genome sequence as a background to understanding the genetics and biology of any given organism.

In many ways genome projects do not confine themselves to only determining a DNA sequence of an organism. Such projects may also include gene prediction to find out where the genes are in a genome, and what those genes do. There may also be related projects to sequence ESTs or mRNAs to help find out where the genes actually are.

Historical and Technological Perspectives

Historically, when sequencing eukaryotic genomes (such as the worm *Caenorhabditis elegans*) it was common to first map the genome to provide a series of landmarks across the genome. Rather than sequence a chromosome in one go, it would be sequenced piece by piece (with the prior knowledge of approximately where that piece is located on the larger chromosome). Changes in technology and in particular improvements to the processing power of computers, means that genomes can now be 'shotgun sequenced' in one go (there are caveats to this approach though when compared to the traditional approach).

Improvements in DNA sequencing technology has meant that the cost of sequencing a new genome sequence has steadily fallen (in terms of cost per base pair) and newer technology has also meant that genomes can be sequenced far more quickly.

When research agencies decide what new genomes to sequence, the emphasis has been on species which are either high importance as model organism or have a relevance to human health (e.g. pathogenic bacteria or vectors of disease such as mosquitos) or species which have commercial importance (e.g. livestock and crop plants). Secondary emphasis is placed on species whose genomes will help answer important questions in molecular evolution (e.g. the common chimpanzee).

In the future, it is likely that it will become even cheaper and quicker to sequence a genome. This will allow for complete genome sequences to be determined from many different individuals of the same species. For humans, this will allow us to better understand aspects of human genetic diversity.

Example Genome Projects

L1 Dominette 01449, the Hereford who serves as the subject of the Bovine Genome Project

Many organisms have genome projects that have either been completed or will be completed shortly, including:

- Humans, Homo sapiens.
- Humans, Homo sapiens.
- Palaeo-Eskimo, an ancient-human.
- Neanderthal, "*Homo neanderthalensis*" (partial).
- Common chimpanzee *Pan troglodytes.*

- Domestic Cow.
- Bovine Genome.
- Honey Bee Genome Sequencing Consortium.
- Horse genome.
- Human microbiome project.
- International Grape Genome Program.
- International HapMap Project.
- Tomato 150+ genome resequencing project.
- 100K Genome Project.
- Genomics England.

Genome-wide Association Study

In genetics, a genome-wide association study (GWA study, or GWAS), also known as whole genome association study (WGA study, or WGAS), is an examination of a genome-wide set of genetic variants in different individuals to see if any variant is associated with a trait. GWASs typically focus on associations between single-nucleotide polymorphisms (SNPs) and traits like major human diseases, but can equally be applied to any other organism.

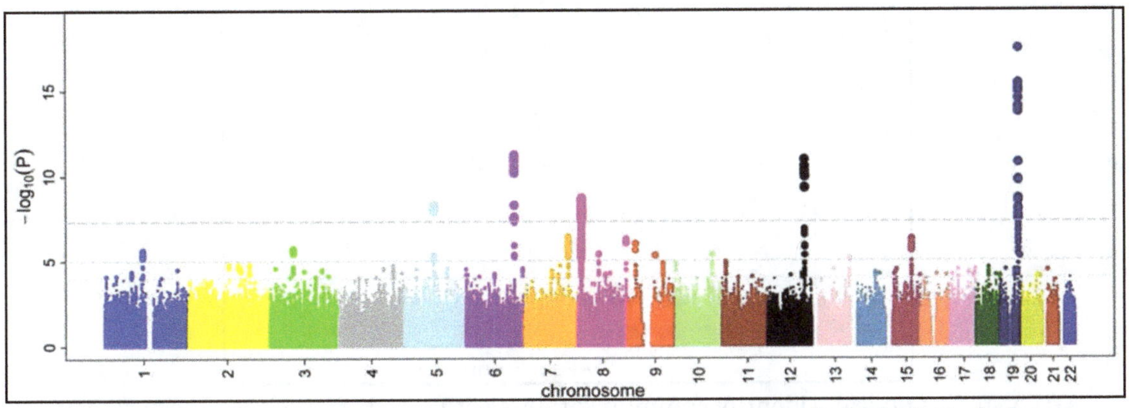

An illustration of a Manhattan plot depicting several strongly associated risk loci. Each dot represents a SNP, with the X-axis showing genomic location and Y-axis showing association level. This example is taken from a GWA study investigating microcirculation, so the tops indicates genetic variants that more often are found in individuals with constrictions in small blood vessels.

In the case of GWAS applied to human data, these studies compare the DNA of participants having varying phenotypes for a particular trait or disease. Participants in a

GWAS study may be people with a disease (cases) and similar people without (controls), or they may be people with different phenotypes for a particular trait, for example blood pressure. This approach is known as phenotype-first, in which the participants are classified first by their clinical manifestation(s), as opposed to genotype-first. Each person gives a sample of DNA, from which millions of genetic variants are read using SNP arrays. If one type of the variant (one allele) is more frequent in people with the disease, the variant is said to be *associated* with the disease. The associated SNPs are then considered to mark a region of the human genome that may influence the risk of disease. In contrast to methods that specifically test one or a few genetic regions, the GWA studies investigate the entire genome. The approach is therefore said to be *non-candidate-driven* in contrast to *gene-specific candidate-driven studies.* GWA studies identify SNPs and other variants in DNA associated with a disease, but they cannot on their own specify which genes are causal.

The first successful GWAS was published in 2005. It investigated patients with age-related macular degeneration and found two SNPs with significantly altered allele frequency compared to healthy controls. As of 2011, hundreds or thousands of individuals are tested, over 1,200 human GWA studies have examined over 200 diseases and traits, and almost 4,000 SNP associations have been found. Several GWA studies have received criticism for omitting important quality control steps, rendering the findings invalid, but modern publications address these issues. However, the methodology itself still has opponents.

Background

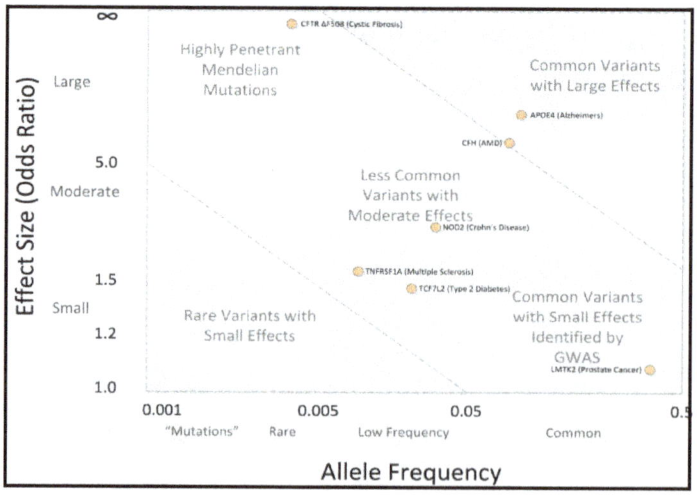

GWA studies typically identify common variants with small effect sizes (*lower right*).

Any two human genomes differ in millions of different ways. There are small variations in the individual nucleotides of the genomes (SNPs) as well as many larger variations, such as deletions, insertions and copy number variations. Any of these may cause alterations in an individual's traits, or phenotype, which can be anything from disease risk to physical properties such as height. Around the year 2000, prior to the introduction of GWA studies, the primary method of investigation was through inheritance studies of genetic linkage in families. This approach had proven highly useful towards single gene disorders. However, for common and complex diseases the results of genetic linkage studies proved hard to reproduce. A suggested alternative to linkage studies was the genetic association study. This study type asks if the allele of a genetic variant is found more often than expected in individuals with the phenotype of interest (e.g. with the disease being studied). Early calculations on statistical power indicated that this approach could be better than linkage studies at detecting weak genetic effects.

In addition to the conceptual framework several additional factors enabled the GWA studies. One was the advent of biobanks, which are repositories of human genetic material that greatly reduced the cost and difficulty of collecting sufficient numbers of biological specimens for study. Another was the International HapMap Project, which, from 2003 identified a majority of the common SNPs interrogated in a GWA study. The haploblock structure identified by HapMap project also allowed the focus on the subset of SNPs that would describe most of the variation. Also the development of the methods to genotype all these SNPs using genotyping arrays was an important prerequisite.

Methods

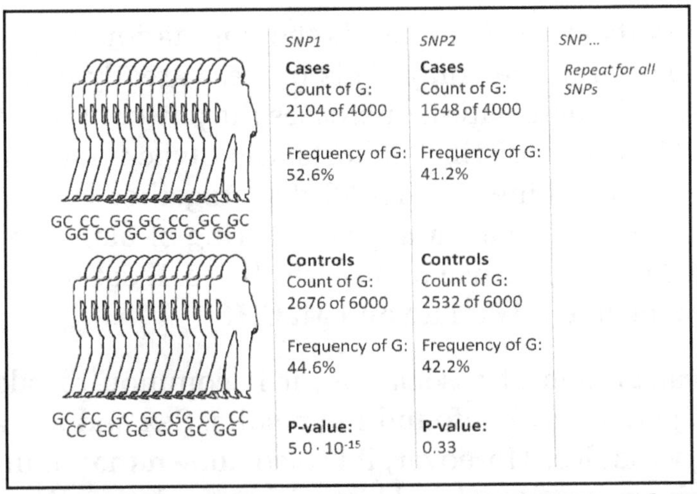

Example calculation illustrating the methodology of a case-control GWA study. The allele count of each measured SNP is evaluated—in this case with a chi-squared test—to identify variants associated with the trait in question. The numbers in this example are taken from a 2007 study of coronary artery disease (CAD) that showed that the individuals with the G-allele of SNP1 (*rs1333049*) were overrepresented amongst CAD-patients.

The most common approach of GWA studies is the case-control setup, which compares two large groups of individuals, one healthy control group and one case group affected by a disease. All individuals in each group are genotyped for the majority of common known SNPs. The exact number of SNPs depends on the genotyping technology, but are typically one million or more. For each of these SNPs it is then investigated if the allele frequency is significantly altered between the case and the control group. In such setups, the fundamental unit for reporting effect sizes is the odds ratio. The odds ratio is the ratio of two odds, which in the context of GWA studies are the odds of disease for individuals having a specific allele and the odds of disease for individuals who do not have that same allele. When the allele frequency in the case group is much higher than in the control group, the odds ratio is higher than 1, and vice versa for lower allele frequency. Additionally, a P-value for the significance of the odds ratio is typically calculated using a simple chi-squared test. Finding odds ratios that are significantly different from 1 is the objective of the GWA study because this shows that a SNP is associated with disease.

There are several variations to this case-control approach. A common alternative to case-control GWA studies is the analysis of quantitative phenotypic data, e.g. height or biomarker concentrations or even gene expression. Likewise, alternative statistics designed for dominance or recessive penetrance patterns can be used. Calculations are typically done using bioinformatics software such as SNPTEST and PLINK, which also include support for many of these alternative statistics. Earlier GWAS focused on the effect of individual SNPs. However, the empirical evidence shows that complex interactions among two or more SNPs, epistasis, might contribute to complex diseases. Moreover, the researchers tries to integrate GWA data with other biological data such as protein protein interaction network to extract more informative results.

A key step in the majority of GWA studies is the imputation of genotypes at SNPs not on the genotype chip used in the study. This process greatly increases the number of SNPs that can be tested for association, increases the power of the study, and facilitates meta-analysis of GWAS across distinct cohorts. Genotype imputation is carried out by statistical methods that combine the GWAS data together with a reference panel of haplotypes. These methods take advantage of sharing of haplotypes between individuals over short stretches of sequence to impute alleles. Existing software packages for genotype imputation include IMPUTE2 and MaCH.

In addition to the calculation of association, it is common to take into account any variables that could potentially confound the results. Sex and age are common examples of confounding variables. Moreover, it is also known that many genetic variations are associated with the geographical and historical populations in which the mutations first arose. Because of this association, studies must take account of the geographical and ethnical background of participants by controlling for what is called population stratification.

After odds ratios and P-values have been calculated for all SNPs, a common approach is to create a Manhattan plot. In the context of GWA studies, this plot shows the negative logarithm of the P-value as a function of genomic location. Thus the SNPs with the most significant association stands out on the plot, usually as stacks of points because of haploblock structure. Importantly, the P-value threshold for significance is corrected for multiple testing issues. The exact threshold varies by study, but the conventional threshold is 5×10^{-8} to be significant in the face of hundreds of thousands to millions of tested SNPs. GWA studies typically perform the first analysis in a discovery cohort, followed by validation of the most significant SNPs in an independent validation cohort.

Results

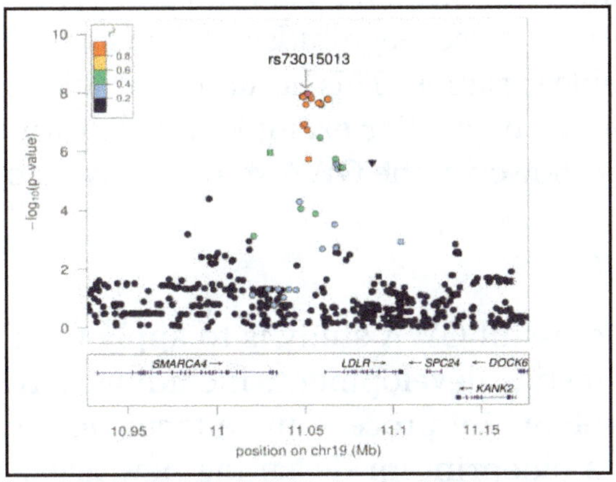

Regional association plot, showing individual SNPs in the LDL receptor region and their association to LDL-cholesterol levels. This type of plot is similar to the Manhattan plot in the lead section, but for a more limited section of the genome. The haploblock structure is visualized with colour scale and the association level is given by the left Y-axis. The dot representing the rs73015013 SNP (in the top-middle) has a high Y-axis location because this SNP explains some of the variation in LDL-cholesterol.

Attempts have been made at creating comprehensive catalogues of SNPs that have been identified from GWA studies. As of 2009, SNPs associated with diseases are numbered in the thousands.

The first GWA study, conducted in 2005, compared 96 patients with age-related macular degeneration (ARMD) with 50 healthy controls. It identified two SNPs with significantly altered allele frequency between the two groups. These SNPs were located in the gene encoding complement factor H, which was an unexpected finding in the research of ARMD. The findings from these first GWA studies have subsequently prompted further functional research towards therapeutical manipulation of the complement system in ARMD. Another landmark publication in the history of GWA studies was the Wellcome Trust Case Control Consortium (WTCCC) study, the largest GWA study ever conducted at the time of its publication in 2007. The WTCCC included 14,000 cases of seven common diseases (~2,000 individuals for each of coronary heart disease, type 1 diabetes, type 2 diabetes, rheumatoid arthritis, Crohn's disease, bipolar disorder, and hypertension) and 3,000 shared controls. This study was successful in uncovering many new disease genes underlying these diseases.

Since these first landmark GWA studies, there have been two general trends. One has been towards larger and larger sample sizes. At the end of 2011, the largest sample sizes were in the range of 200,000 individuals. The reason is the drive towards reliably detecting risk-SNPs that have smaller odds ratios and lower allele frequency. Another trend has been towards the use of more narrowly defined phenotypes, such as blood lipids, proinsulin or similar biomarkers. These are called *intermediate phenotypes*, and their analyses may be of value to functional research into biomarkers.

A central point of debate on GWA studies has been that most of the SNP variations found by GWA studies are associated with only a small increased risk of the disease, and have only a small predictive value. The median odds ratio is 1.33 per risk-SNP, with only a few showing odds ratios above 3.0. These magnitudes are considered small because they do not explain much of the heritable variation. This heritable variation is known from heritability studies based on monozygotic twins. For example, it is known that 80–90% of height is heritable. Of these 80-90%, however, the GWA studies only account for a minority.

Clinical Applications

A challenge for future successful GWA study is to apply the findings in a way that accelerates drug and diagnostics development, including better integration of genetic studies into the drug-development process and a focus on the role of genetic variation in maintaining health as a blueprint for designing new drugs and diagnostics. Several studies have looked into the use of risk-SNP markers as a means of directly improving the accuracy of prognosis. Some have found that the accuracy of prognosis improves, while others report only minor benefits from this use. Generally, a problem with this direct approach is the small magnitudes of the effects observed. A small effect ultimately translates into a poor separation of cases and controls and thus only a small improvement of prognosis accuracy. An alternative application is therefore the potential for GWA studies to elucidate pathophysiology.

One such success is related to identifying the genetic variant associated with response to anti-hepatitis C virus treatment. For genotype 1 hepatitis C treated with Pegylated interferon-alpha-2a or Pegylated interferon-alpha-2b combined with ribavirin, a GWA study has shown that SNPs near the human IL28B gene, encoding interferon lambda 3, are associated with significant differences in response to the treatment. A later report demonstrated that the same genetic variants are also associated with the natural clearance of the genotype 1 hepatitis C virus. These major findings facilitated the development of personalized medicine and allowed physicians to customize medical decisions based on the patient's genotype.

The goal of elucidating pathophysiology has also led to increased interest in the association between risk-SNPs and the gene expression of nearby genes, the so-called expression quantitative trait loci (eQTL) studies. The reason is that GWAS studies identify risk-SNPs, but not risk-genes, and specification of genes is one step closer towards actionable drug targets. As a result, major GWA studies of 2011 typically included extensive eQTL analysis. One of the strongest eQTL effects observed for a GWA-identified risk SNP is the SORT1 locus. Functional follow up studies of this locus using small interfering RNA and gene knock-out mice have shed light on the metabolism of low-density lipoproteins, which have important clinical implications for cardiovascular disease.

Limitations

GWA studies have several issues and limitations that can be taken care of through proper quality control and study setup. Lack of well defined case and control groups, insufficient sample size, control for multiple testing and control for population stratification are common problems. Particularly the statistical issue of multiple testing wherein it has been noted that "the GWA approach can be problematic because the massive number of statistical tests performed presents an unprecedented potential for false-positive results". Ignoring these correctible issues has been cited as contributing to a general sense of problems with the GWA methodology. In addition to easily correctible problems such as these, some more subtle but important issues have surfaced. A high-profile GWA study that investigated individuals with very long life spans to identify SNPs associated with longevity is an example of this. The publication came under scrutiny because of a discrepancy between the type of genotyping array in the case and control group, which caused several SNPs to be falsely highlighted as associated with longevity. The study was subsequently retracted.

In addition to these preventable issues, GWA studies have attracted more fundamental criticism, mainly because of their assumption that common genetic variation plays a large role in explaining the heritable variation of common disease. This aspect of GWA studies has attracted the criticism that, although it could not have been known prospectively, GWA studies were ultimately not worth the expenditure. Alternative strategies suggested involve linkage analysis.More recently, the rapidly decreasing price of complete genome sequencing have also provided a realistic alternative to genotyping array-based GWA studies. It can be discussed if the use of this new technique is still referred to as a GWA study, but high-throughput sequencing does have potential to side-step some of the shortcomings of non-sequencing GWA.

Fine-mapping

Genotyping arrays designed for GWAS rely on linkage disequilibrium to provide coverage of the entire genome by genotyping a subset of variants. Because of this, the reported associated variants are unlikely to be the actual causal variants. Associated regions can contain hundreds of variants spanning large regions and encompassing many different genes, making the biological interpretation of GWAS loci more difficult. Fine-mapping is a process to refine these lists of associated variants to a credible set most likely to include the causal variant.

Fine-mapping requires all variants in the associated region to have been genotyped or imputed (dense coverage), very stringent quality control resulting in high-quality genotypes, and large sample sizes sufficient in separating out highly correlated signals. There are several different methods to perform fine-mapping, and all methods produce a posterior probability that a variant in that locus is causal. Because the requirements are often difficult to satisfy, there are still limited examples of these methods being more generally applied.

References

- Bergman NH, ed. (2007). Comparative Genomics: Volumes 1 and 2. Totowa (NJ): Humana Press. ISBN 978-193411-537-4. PMID 21250292
- Herrero A, Flores E, eds. (2008). The Cyanobacteria: Molecular Biology, Genomics and Evolution (1st ed.). Caister Academic Press. ISBN 978-1-904455-15-8
- Marco D, ed. (2011). Metagenomics: Current Innovations and Future Trends. Caister Academic Press. ISBN 978-1-904455-87-5
- McElheny V (2010). Drawing the map of life : inside the Human Genome Project. New York NY: Basic Books. ISBN 9780465043330
- McGrath S, van Sinderen D, eds. (2007). Bacteriophage: Genetics and Molecular Biology (1st ed.). Caister Academic Press. ISBN 978-1-904455-14-1
- Keith JM, ed. (2008). "Bioinformatics". Methods in Molecular Biology. 453. doi:10.1007/978-1-60327-429-6. ISBN 978-1-60327-428-9
- Church GM, Regis E (2012). Regenesis : how synthetic biology will reinvent nature and ourselves. New York: Basic Books. ISBN 9780465021758

Chapter 4

Algorithms in Bioinformatics

An algorithm is a set of rules or procedures to be followed by a computer for solving problems. Bioinformatics makes use of algorithms which have been adapted for application in solving problems related to molecular and genome biology. Baum-Welch algorithm, Needleman-Wunsch algorithm, Hirschberg's algorithm, etc. are a few examples of bioinformatics algorithms which are discussed thoroughly in this chapter.

Baum–Welch Algorithm

A cornerstone of stochastic grammar-based profiling of DNA, protein and RNA sequences is the fast and accurate estimation of the probability parameters of the model, given a 'training set'. This is achieved for hidden Markov models (HMMs) using the Baum–Welch algorithm.

Let us consider discrete (categorical) HMMs of length T (each observation sequence is T observations long). Let the space of observations be $X = \{1, 2, ..., N\}$, and let the space of underlying states be $Z = \{1, 2, ..., M\}$. An HMM $\theta = (\pi, A, B)$ is parameterized by the initial state matrix π, the state transition matrix A, and the emission matrix B; $\pi_i = P(z_1 = i)$, $A_{ij} = P(z_{t+1} = j \mid z_t = 1)$, and $B_i(j) = P(x_t = j \mid zt = i)$.

We study the problem of learning the parameterization of θ from a dataset of D observations. Let $\chi = \left(X^{(1)}, ..., X^{(D)}\right)$, where each $X^{(i)} = \left(x_1^{(i)}, x_2^{(i)}, ..., x_T^{(i)}\right)$ We assume each observation is drawn iid. The learning problem is nontrivial because we are not given the latent variable $Z^{(i)}$ for each $X^{(i)}$, otherwise we could directly compute $\theta^* = argmax_\theta\ P(\mathcal{X}, \mathcal{Z}; \theta)$ Without $\mathcal{Z}$, the naive solution would be to directly compute $\theta^* = \text{argmax}_\theta \sum_{z \in \mathcal{Z}} P(\mathcal{X}, z; \theta)$ This is not tractable, since there are DT^M different values of z to try.

Baum-Welch is an iterative procedure for estimating θ^* from only $\mathcal{X}$. It works by maximizing a proxy to the log-likelihood, and updating the current model to be closer to the optimal model. Each iteration of Baum-Welch is guaranteed to increase the log-likelihood of the data. But of course, convergence to the optimal solution is not guaranteed. Baum-Welch can be described simply as repeating the following steps until convergence:

1. Compute $Q(\theta, \theta^s) = \sum_{z \in \mathcal{Z}} \log \left[P(\mathcal{X}, z; \theta)\right] P(z \mid X; \theta^s)$.
2. Set $\theta^{s+1} = \underset{\theta}{\text{argmax}}\ Q(\theta, \theta^s)$.

Without justifying why this works, the rest of this document will focus on deriving the necessary update steps to run this algorithm. First, noting that $P(z, \mathcal{X}) = P(\mathcal{X})P(z \mid \mathcal{X})$, we can write

$$\underset{\theta}{\operatorname{argmax}} \sum_{z \in Z} \log \left[P(\mathcal{X}, z; \theta)\right] P(z \mid \mathcal{X}; \theta^{s}) = \underset{\theta}{\operatorname{argmax}} \sum_{z \in Z} \log \left[P(\mathcal{X}, z; \theta)\right] P(z \mid \mathcal{X}; \theta^{s}) = \underset{\theta}{\operatorname{argmax}}\ \hat{Q}(\theta, \theta^{s})$$

since $P(\mathcal{X})$ is not affected by choice of θ. Now $P(z, \mathcal{X}; \theta)$ is easy to write down

$$P(z, \mathcal{X}; \theta) = \prod_{d=1}^{D} \left(\pi_{z_1^{(d)}} B_{z_1^{(d)}}\left(x_1^{(d)}\right) \prod_{t=2}^{T} A_{z_{t-1}^{(d)} z_t^{(d)}} B_{z_t^{(d)}}\left(x_t^{(d)}\right) \right)$$

Taking the log gives us

$$\log P(z, \mathcal{X}; \theta) = \sum_{d=1}^{D} \left[\log \pi_{z_1^{(d)}} + \sum_{t=2}^{T} \log A_{z_{t-1}^{(d)} z_t^{(d)}} + \sum_{t=1}^{T} \log B_{z_t^{(d)}}\left(x_t^{(d)}\right) \right]$$

Plugging this into $\hat{Q}(\theta, \theta^{s})$, we get

$$\hat{Q}(\theta, \theta^{s}) = \sum_{z \in Z} \sum_{d=1}^{D} \log \pi_{z_1^{(d)}} P(z, \mathcal{X}; \theta^{s}) + \sum_{z \in Z} \sum_{d=1}^{D} \sum_{t=2}^{T} \log A_{z_{t-1}^{(d)} z_t^{(d)}} P(z, \mathcal{X}; \theta^{s}) + \sum_{z \in Z} \sum_{d=1}^{D} \sum_{t=1}^{T} \log B_{z_t^{(d)}}\left(x_t^{(d)}\right) P(z, \mathcal{X}; \theta^{s})$$

This is a nice form which we can optimize analytically with Lagrange multipliers. We need Lagrange multipliers because we have equality constraints which come from requiring that π, $A_i\cdot$ and $B_i(\cdot)$ form valid probability distributions. Let $\hat{L}(\theta, \theta^{s})$ be the Lagrangian

$$\hat{L}(\theta, \theta^{s}) = \hat{Q}(\theta, \theta^{s}) - \lambda_{\pi}\left(\sum_{i=1}^{M} \pi_i - 1\right) - \sum_{i=1}^{M} \lambda_{A_i}\left(\sum_{i=1}^{M} A_{ij} - 1\right) - \sum_{i=1}^{M} \lambda_{B_i}\left(\sum_{i=1}^{N} B_i(j) - 1\right)$$

First let us focus on the π_i 's

$$\begin{aligned}
\frac{\partial \hat{L}(\theta, \theta^{s})}{\partial \pi_i} &= \frac{\partial}{\partial \pi_i}\left(\sum_{z \in Z} \sum_{d=1}^{D} \log \pi_{z_1^{(d)}} P(z, X; \theta^{s})\right) - \lambda_{\pi} = 0 \\
&= \frac{\partial}{\partial \pi_i}\left(\sum_{j=1}^{M} \sum_{d=1}^{D} \log \pi_j P\left(z_1^{(d)} = j, \mathcal{X}; \theta^{s}\right)\right) - \lambda_{\pi} = 0 \\
&= \sum_{d=1}^{D} \frac{P\left(z_1^{(d)} = i, \mathcal{X}; \theta^{s}\right)}{\pi_i} - \lambda_{\pi} = 0 \\
\frac{\partial \hat{L}(\theta, \theta^{s})}{\partial \lambda_{\pi}} &= -\left(\sum_{j=1}^{M} \pi_i - 1\right) = 0
\end{aligned}$$

The second step is simply the result of marginalizing out, for each d, all $z_{t \neq 1}^{(d)}$ and $z_t^{(d' \neq d)}$ for all t . Some algebra yields

$$\pi_i = \frac{\sum_{d=1}^{D} P\left(z_1^{(d)} = i, X; \theta^{s}\right)}{\sum_{j=1}^{M} \sum_{d=1}^{D} P\left(z_1^{(d)} = j, X; \theta^{s}\right)} = \frac{\sum_{d=1}^{D} P\left(z_1^{(d)} = i, X; \theta^{s}\right)}{\sum_{d=1}^{M} \sum_{j=1}^{D} P\left(z_1^{(d)} = j, X; \theta^{s}\right)}$$

$$=\frac{\sum_{d=1}^{D} P\left(z_1^{(d)}=i, \mathcal{X} \; ; \theta^s\right)}{\sum_{d=1}^{D} P\left(\mathcal{X} \; ; \theta^s\right)}=\frac{\sum_{d=1}^{D} P\left(z_1^{(d)}=i, \mathcal{X} \; ; \theta^s\right)}{D\, P\left(\mathcal{X} \; ; \theta^s\right)}$$

$$=\frac{\sum_{d=1}^{D} P\left(\mathcal{X} \; ; \theta^s\right) P\left(z_1^{(d)}=i \mid \mathcal{X} \; ; \theta^s\right)}{D\, P\left(\mathcal{X} \; ; \theta^s\right)}=\frac{1}{D} \sum_{d=1}^{D} P\left(z_1^{(d)}=i \mid \mathcal{X} \; ; \theta^s\right)$$

$$=\frac{1}{D} \sum_{d=1}^{D} P\left(z_1^{(d)}=i \mid X^{(d)} \; ; \theta^s\right)$$

We now follow a similar process for the A_{ij} 's.

$$\frac{\partial \hat{L}\left(\theta, \theta^s\right)}{\partial A_{ij}}=\frac{\partial}{\partial A_{ij}}\left(\sum_{z \in Z} \sum_{d=1}^{D} \sum_{t=2}^{T} \log A_{z_{t-1}^{(d)} z_t^{(d)}} P\left(z, \mathcal{X} \; ; \theta^s\right)\right)-\lambda_{A_i}=0$$

$$=\frac{\partial}{\partial A_{ij}}\left(\sum_{j=1}^{M} \sum_{k=1}^{M} \sum_{d=1}^{D} \sum_{t=2}^{T} \log A_{jk} P\left(z_{t-1}^{(d)}=j, z_t^{(d)}=k, \mathcal{X} \; ; \theta^s\right)\right)-\lambda_{A_i}=0$$

$$=\sum_{d=1}^{D} \sum_{t=2}^{T} \frac{P\left(z_{t-1}^{(d)}=i, z_t^{(d)}=j, \mathcal{X} \; ; \theta^s\right)}{A_{ij}}-\lambda_{A_i}=0$$

$$\frac{\partial \hat{L}\left(\theta, \theta^s\right)}{\partial A_i}=-\left(\sum_{j=1}^{M} A_{ij}-1\right)=0$$

This yields

$$A_{ij}=\frac{\sum_{d=1}^{D} \sum_{t=2}^{T} P\left(z_{t-1}^{(d)}=i, z_t^{(d)}=j, \mathcal{X} \; ; \theta^s\right)}{\sum_{j=1}^{M} \sum_{d=1}^{D} \sum_{t=2}^{T} P\left(z_{t-1}^{(d)}=i, z_t^{(d)}=j, \mathcal{X} \; ; \theta^s\right)}$$

$$=\frac{\sum_{d=1}^{D} \sum_{t=2}^{T} P\left(z_{t-1}^{(d)}=i, z_t^{(d)}=j, \mathcal{X} \; ; \theta^s\right)}{\sum_{d=1}^{D} \sum_{t=2}^{T} P\left(z_{t-1}^{(d)}=i, \mathcal{X} \; ; \theta^s\right)}$$

$$=\frac{\sum_{d=1}^{D} \sum_{t=2}^{T} P\left(\mathcal{X} \; ; \theta^s\right) P\left(z_{t-1}^{(d)}=i, z_t^{(d)}=j \mid \mathcal{X} \; ; \theta^s\right)}{\sum_{d=1}^{D} \sum_{t=2}^{T} P\left(\mathcal{X} \; ; \theta^s\right) P\left(z_{t-1}^{(d)}=i \mid \mathcal{X} \; ; \theta^s\right)}$$

$$=\frac{\sum_{d=1}^{D} \sum_{t=2}^{T} P\left(z_{t-1}^{(d)}=i, z_t^{(d)}=j \mid \mathcal{X}^{(d)} \; ; \theta^s\right)}{\sum_{d=1}^{D} \sum_{t=2}^{T} P\left(z_{t-1}^{(d)}=i \mid \mathcal{X}^{(d)} ; \theta^s\right)}$$

The final thing is the $B_i(j)$'s, which are slightly trickier. Let $I(x)$ denote an indicator function which is 1 if x is true, 0 otherwise.

$$\frac{\partial \hat{L}(\theta, \theta^s)}{\partial B_i(j)} = \frac{\partial}{\partial B_i(j)}\left(\sum_{z \in Z}\sum_{d=1}^{D}\sum_{t=1}^{T}\log B_{z_t^{(d)}}\left(x_t^{(d)}\right)P\left(z, \mathcal{X} ; \theta^s\right)\right) - \lambda_{B_i} = 0$$

$$= \frac{\partial}{\partial B_i(j)}\left(\sum_{i=1}^{N}\sum_{d=1}^{D}\sum_{t=1}^{T}\log B_i\left(x_t^{(d)}\right)P\left(z_t^{(d)} = i, \mathcal{X} ; \theta^s\right)\right) - \lambda_{B_i} = 0$$

$$= \sum_{d=1}^{D}\sum_{t=1}^{T}\frac{P\left(z_t^{(d)} = i, \mathcal{X} ; \theta^s\right)I\left(x_t^{(d)} = j\right)}{B_i(j)} - \lambda_{B_i} = 0$$

$$\frac{\partial \hat{L}(\theta, \theta^s)}{\partial \lambda_{B_i}} = -\left(\sum_{j=1}^{N} B_i(j) - 1\right) = 0$$

This should come as no surprise by now

$$B_i(j) = \frac{\sum_{d=1}^{D}\sum_{t=1}^{T}P\left(z_t^{(d)} = i, \mathcal{X} ; \theta^s\right)I\left(x_t^{(d)} = j\right)}{\sum_{i=1}^{N}\sum_{d=1}^{D}\sum_{t=1}^{T}P\left(z_t^{d} = i, \mathcal{X} ; \theta^s\right)I\left(x_t^{(d)} = j\right)}$$

$$= \frac{\sum_{d=1}^{D}\sum_{t=1}^{T}P\left(z_t^{(d)} = i, \mathcal{X} ; \theta^s\right)I\left(x_t^{(d)} = j\right)}{\sum_{d=1}^{D}\sum_{t=1}^{T}P\left(z_t^{(d)} = i, \mathcal{X}^{(d)} ; \theta^s\right)}$$

$$= \frac{\sum_{d=1}^{D}\sum_{t=1}^{T}P\left(z_t^{(d)} = i \mid X^{(d)} ; \theta^s\right)I\left(x_t^{(d)} = j\right)}{\sum_{d=1}^{D}\sum_{t=1}^{T}P\left(z_t^{(d)} = i \mid X^{(d)} ; \theta^s\right)}$$

To summarize, the update steps are

$$\pi_i^{(s+1)} = \frac{1}{D}\sum_{d=1}^{D}P\left(z_t^{(d)} = i \mid X^{(d)} ; \theta^s\right)$$

$$A_{ij}^{(s+1)} = \frac{\sum_{d=1}^{D}\sum_{t=2}^{T}P\left(z_{t-1}^{(d)} = i, z_t^{(d)} = j \mid X^{(d)} ; \theta^s\right)}{\sum_{d=1}^{D}\sum_{t=2}^{T}P\left(z_{t-1}^{(d)} = i, \mid X^{(d)} ; \theta^s\right)}$$

$$B_i^{(s+1)}(j) = \frac{\sum_{d=1}^{D}\sum_{t=1}^{T}P\left(z_t^{(d)} = i \mid X^{(d)} ; \theta^s\right)I\left(x_t^{(d)} = j\right)}{\sum_{d=1}^{D}\sum_{t=1}^{T}P\left(z_t^{(d)} = i \mid X^{(d)} ; \theta^s\right)}$$

Note that $P(z_t \mid X;\, \theta)$ and $P(z_{t-1},\, z_t \mid X;\, \theta)$ are both quantities which can be computed efficiently for HMMs by the forward-backwards algorithm.

Needleman–Wunsch Algorithm

Dynamic programming algorithms find the best solution by breaking the original problem into smaller sub-problems and then solving. The Needleman-Wunsch algorithm is a dynamic programming algorithm for optimal sequence alignment. Basically, the concept behind the Needleman-Wunsch algorithm stems from the observation that any partial sub-path that tends at a point along the true optimal path must itself be the optimal path leading up to that point. Therefore the optimal path can be determined by incremental extension of the optimal sub-paths. In a Needleman-Wunsch alignment, the optimal path must stretch from beginning to end in both sequences.

Working of Needleman -Wunsch Algorithm

To study the algorithm, consider the two given sequences.

CGTGAATTCAT (sequence 1), GACTTAC (sequence 2)

The length (count of the nucleotides or amino acids) of the sequence 1 and sequence 2 are 11 and 7 respectively. The initial matrix is created with A+1 column's and B+1 row's (where A and B corresponds to length of the sequences). Extra row and column is given, so as to align with gap, at the starting of the matrix as shown in figure.

	-	C	G	T	G	A	A	T	T	C	A	T
-												
G												
A												
C												
T												
T												
A												
C												

Figure: Initial matrix

After creating the initial matrix, scoring schema has to be introduced which can be user defined with specific scores. The simple basic scoring schema can be assumed as, if two residues (nucleotide or amino acid) at i^{th} and j^{th} position are same, matching score is 1 $\left(S(i,j)=\ 1\right)$ or if the two residues at i^{th} and j^{th} position are not same, mismatch score is assumed as -1 $\left(S(i,j)=\ -1\ \right)$. The gap score (w) or gap penalty is assumed as -1 .

Gap score is defined as penalty given to alignment, when we have insertion or deletion.

The dynamic programming matrix is defined with three different steps:

1. Initialization of the matrix with the scores possible.

2. Matrix filling with maximum scores.

3. Trace back the residues for appropriate alignment.

Initialization Step

This example assumes that there is gap penalty. First row and first column of the matrix can be initially filled with 0. If the gap score is assumed, the gap score can be added to the previous cell of the row or column.

	-	C	G	T	G	A	A	T	T	C	A	T
-	0	-1	-2	-3	-4	-5	-6	-7	-8	-9	-10	-11
G	-1											
A	-2											
C	-3											
T	-4											
T	-5											
A	-6											
C	-7											

Initialization of matrix

Matrix Fill Step

The second and crucial step of the algorithm is matrix filling starting from the upper left hand corner of the matrix. To find the maximum score of each cell, it is required to know the neighbouring scores (diagonal, left and right) of the current position. From the assumed values, add the match or mismatch (assumed) score to the diagonal value. Similarly add the gap score to the other neighbouring values. Thus, we can obtain three different values, from that take the maximum among them and fill the ith and j^{th} position with the score obtained.

In terms of matrix positions, it is important to know $[M_{(i-1,j-1)} + S_{(i,j)}, M_{(i,j-1)} + w, M_{(i-1,j)} + w]$

Overall the equation can be showed in the following manner

$$M_{i,j} = \text{Maximum } [M_{i-1,j-1} + S_{i,j}, M_{i,j-1} + w, M_{i-1,j} + w]$$

To score the matrix of the current position (the first position $M_{1,1}$) the above stated formulae can be used. The first residue (nucleotides or amino acids) in the 2 sequences are 'G' and 'C'. Since they are mismatching residues, the score would ($S_{i,j} = -1$) be -1.

$$\begin{aligned} M_{1,1} &= Max\left[M_{0,0} + S_{1,1}, M_{1,0} + w, M_{0,1} + w\right] \\ &= Max\left[0 + (-1), 0 + (-1), 0 + (-1)\right] \\ &= Max\left[-1, -1, -1, \right] \\ &= -1 \end{aligned}$$

The obtained score -1 is placed in position i, j $(1,1)$ of the scoring matrix. Similarly using the above equation and method, fill all the remaining rows and columns. Place the back pointers to the cell from where the maximum score is obtained, which are predecessors of the current cell.

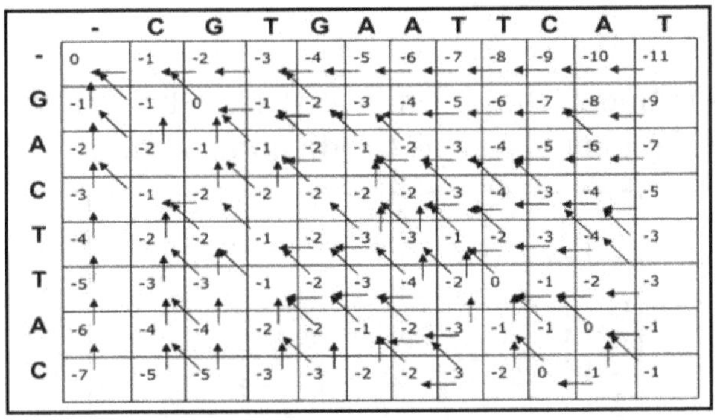

Matrix filling with back pointers

Trace back Step

The final step in the algorithm is the trace back for the best alignment. In the above mentioned example, one can see the bottom right hand corner score as -1. The important point to be noted here is that there may be two or more alignments possible between the two example sequences. The current cell with value -1 has immediate predecessor, where the maximum score obtained is diagonally located and its value is 0. If there are two or more values which points back, suggests that there can be two or more possible alignments.

By continuing the trace back step by the above defined method, one would reach to the 0th row, 0th column. Following the above described steps, alignment of two sample sequences can be found. The best alignment among the alignments can be identified by using the maximum alignment score (match =5, mismatch=-1, gap=-2) which may be user defined.

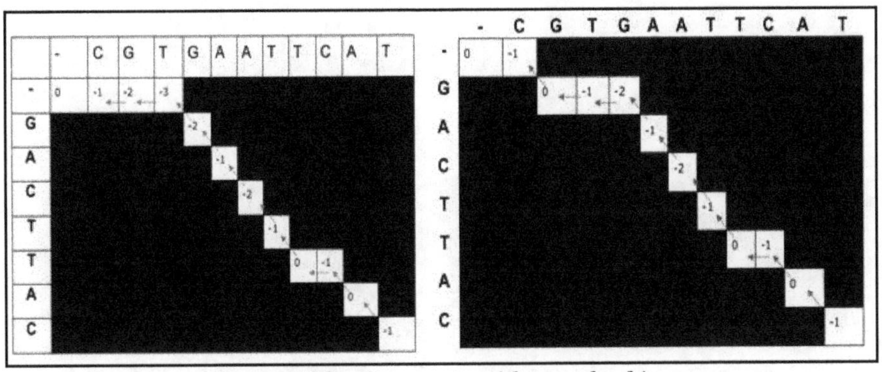

The possible Alignments with trace backing

Hirschberg's Algorithm

While the Needleman-Wunsch algorithm works well for sequence alignment, its space complexity (O(mn)) limits the size of sequences it can align. Hirschberg's algorithm uses a divide and conquer strategy to decrease the space requirement. Specifically, for two sequences (m and n) the first string is cut (m1 and m2) and the second string is cut in a corresponding place (into n1 and n2). The alignment is then solved recursively on m1 and n1, and m2 and n2. It is important to note that the two sub-strings (i.e. n1 and n2) need not have the same length.

Example:

With respect to sequence alignment, the objective is to take the sum at given intervals and use the value corresponding to the alignment score as the dividing point.

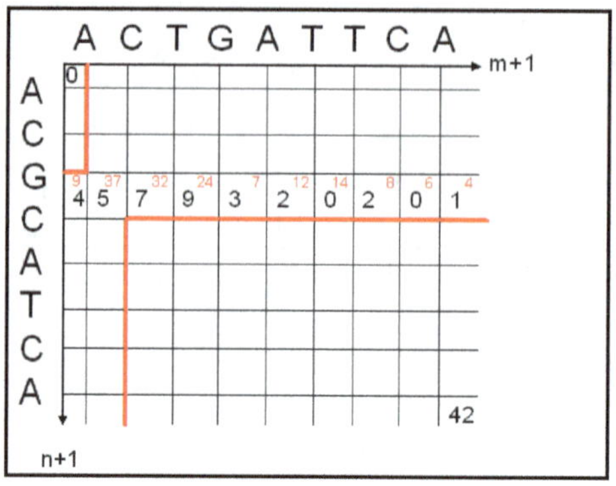

In this figure the sum up to the dividing line (37 + 5) adds up to the alignment score

Example:

What is the sequence of recursive calls on the following sequence alignment when using Hirschberg's algorithm:

```
ACTG

ACTT
```

The alignment is:

```
s1: ACTG

     ||| s2: ACTT
```

Solution:

```
s1[0..3] : s2[0..3]

    s1[0..1] : s2[0..1]

        s1[0..0] : s2[0..0]

        s1[1..1] : s2[1..1]

    s1[2..3] : s2[2..3]

        s1[2..2] : s2[2..2]

        s1[3..3] : s2[3..3]
```

Relating Local Alignments to Global Alignments

When aligning two very large sequences, it is often useful to determine the locations of high similarity regions. Now that we know how to calculate the *global* alignments, how can we find all local high-scoring hits, or *local* alignments above a given threshold for two large sequences? The answer is related to a programming "pearl", the 'Maximum Contiguous Subsequence Sum' (MSS).

Problem:

Given integers $A_1, A_2, \ldots, A_N$ find (and identify the sequence corresponding to) the maximum value of:

$$\sum_{k=1}^{j} A_k$$

Solution:

Can be solved in time complexity of 'n'.

```
mss(A) {
    max = 0;
    sum = 0;
            for (i=1; i ≤ n; i+1)
                { sum = sum +
                A[i]; if (sum >
                max)
                        max =
                sum; if (sum <
                0)
                    sum = 0;
  }
        return max;
        }
```

Analysis

When a subsequence occurs which has a negative sum, the subsequence which will be examined next can begin after the first subsequence (the one that produced the negative sum). Basically, the entire first subsequence is regarded as not having a starting point which will generate a positive sum. For example, consider this set of numbers:

4, 6, -2, 2, -14, 9

Some sums are positive $(4,\ 4+6,\ 4+6+(-2), 4+6+(-2)+2)$ but the sum of the first 5 terms $(4+6+(-2)+2-14)$ is negative. Therefore it follows logically that any sequence starting between the 4 and -14 and ending with the -14 will have a negative sum.

The maximum contiguous subsequence sum searches exactly for the highest scoring local area.

Smith–Waterman Algorithm

Over a decade after the initial publication of the Needleman-Wunsch algorithm, a modification was made to allow for local alignments. Today, the Smith-Waterman alignment algorithm is the one used by the Basic Local Alignment Search Tool (BLAST) which is the most cited resource in biomedical literature.

Working of Smith-Waterman Algorithm

Intialization of Matrix

The basic steps for the algorithm are similar to that of Needleman-Wunsch algorithm. The steps are:

1. Initialization of a matrix.

2. Matrix Filling with the appropriate scores.

3. Trace back the sequences for a suitable alignment.

To study the Local sequence alignment consider the given below sequences.

CGTGAATTCAT (sequence 1 or A)

GACTTAC (sequence 2 or B)

The two sequences are arranged in a matrix form with A+1columns and B+1rows. The values in the first row and first column are set to zero as shown in figure.

	-	C	G	T	G	A	A	T	T	C	A	T
-	0	0	0	0	0	0	0	0	0	0	0	0
G	0											
A	0											
C	0											
T	0											
T	0											
A	0											
C	0											

Initialization of Matrix

Variables used:

Variables used:

i, j describes row and columns.

M is the matrix value of the required cell (stated as $M_{i,j}$)

S is the score of the required cell $\left(S_{i,j}\right)$

W is the gap alignment

Matrix Filling

The second and crucial step of the algorithm is filling the entire matrix, so it is more important to know the neighbor values (diagonal, upper and left) of the current cell to fill each and every cell.

$$M_{i,j} = Maximum\ \left[M_{i-1,j-1} + S_{i,j.}\ M_{i,j-1} + W, M_{i-1,j} + W\ , 0\right]$$

As per the assumptions stated earlier, fill the entire matrix using the assumed scoring schema and initial values. One can fill the 1st row and 1st column with the scoring matrix as follows.

The first residue (nucleotides or amino acids) in both sequences is 'C' and 'G', the matching score or the mismatching score is going to be added the neighboring value which is diagonally located i.e. 0. The upper and left values are added to the gap penalty score from the matrix. So the scoring schema equation can be shown as follows.

$$\begin{aligned} M_{1,1} &= Maximum\left[M_{0,0} + S_{1,1}, M_{1,0} + w, M_{0,1} + W, 0\right] \\ &= Maximum\left[0(-3), 0+(-4), 0+(-4), 0\right] \\ &= Maximum\left[-3, -4, -4, 0\right] \\ &= 0 \end{aligned}$$

From the above calculations the maximum value obtained is 0. Finding the maximum value for $M_{i,j}$ position, one can notice that there is no chance to see any negative values in the matrix, since we are taking 0 as lowest value.

After filling the matrix, keep the pointer back to the cell from where the maximum score has been determined. In the similar fashion fill all the values of the matrix of the cell.

For the example the matrix can be filled is shown in figure.

	-	C	G	T	G	A	A	T	T	C	A	T
-	0	0	0	0	0	0	0	0	0	0	0	0
G	0	0	5	1	5	1	0	0	0	0	0	0
A	0	0	1	2	1	10	6	2	0	0	5	1
C	0	5	1	0	0	6	7	3	0	5	1	2
T	0	1	2	6	2	2	3	12	8	4	2	6
T	0	0	0	7	3	0	0	8	17	13	9	7
A	0	0	0	3	4	8	5	4	13	14	18	14
C	0	5	1	0	0	4	5	2	9	18	14	15

Matrix filling with back pointers

Each cell is back pointed by one or more pointers from where the maximum score has been obtained.

Trace Backing the Sequences for an Optimal Alignment

The final step for the appropriate alignment is trace backing, prior to that one needs to find out the maximum score obtained in the entire matrix for the local alignment of the sequences. It is possible that the maximum scores can be present in more than one cell, in that case there may be possibility of two or more alignments, and the best alignment by scoring it.

In this example we can see the maximum score in the matrix as 18, which is found in two positions that lead to multiple alignments, so the best alignment, has to be found.

So the trace back begins from the position which has the highest value, pointing back with the pointers, thus find out the possible predecessor, then move to next predecessor and continue until we reach the score 0.

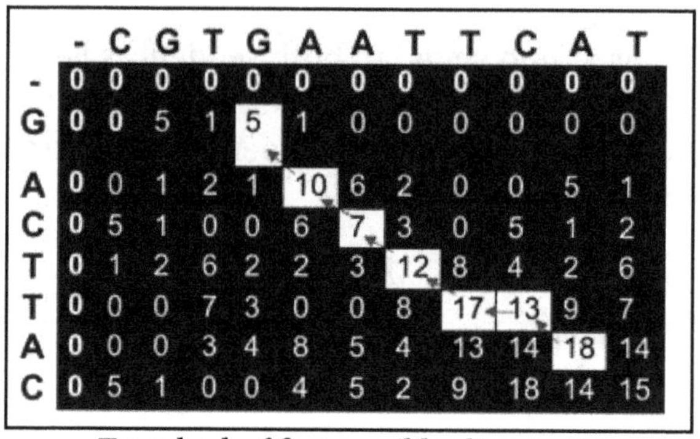

Trace back of first possible alignment

It is possible to find two pointers pointing out from one cell, where both ways (alignments) can be considered, best one is found by scoring and finding maximum score among them.

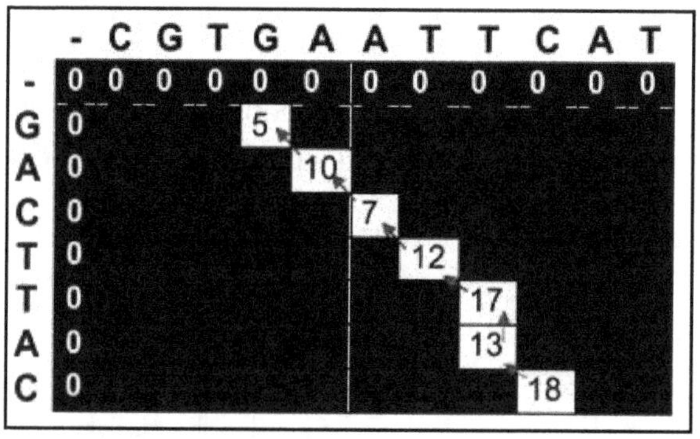

Trace back of second possible alignment

Thus a local alignment is obtained and one can see the possible alignments as in figure.

```
G A A T T C A   G A A T T - C
| | | | |     |   | | | | |     |
G A C T T - A   G A C T T A C

+ + - + + - +   + + - + + - +
5 5 3 5 5 4 5   5 5 3 5 5 4 5
```

Scoring for best alignment

The two alignments can be given with a score, for matching as +5, mismatch as -3 and gap penalty as -4, sum up all the individual scores and the alignment which has maximum score after this can be taken as the best alignment.

By summing up the scores both of the alignments are giving the same as 18, so one can predict both alignments are the best.

Sequence Clustering

In bioinformatics, sequence clustering algorithms have been used to automatically group large protein datasets into different families, to search for protein sequences that are homologous to a given sequence, and to map or align a given DNA sequence to an entire genome, to cite only some of the most common applications. In all of these applications, sequence clustering becomes a valuable tool to gain insight into otherwise seemingly senseless sequences of data.

A similar kind of challenge arises in process mining, where the goal is to extract meaningful task sequences from an event log, usually resorting to special-purpose algorithms that can recover the original workflow that produced the log.

The idea of applying sequence clustering to process mining comes at a time when process mining is still heavily dependent on the assumption that the event log contains "sufficient" information, i.e., that each event in the log is clearly associated with a specific activity and case (process instance) . This comes as a major disadvantage since the classes of information systems that are able to generate such logs are restricted to process-aware systems, and it becomes impossible to apply and benefit from process mining in scenarios where the log data is not available in that form.

A sequence clustering approach can alleviate these requirements by grouping similar sequences and identifying typical ones without the need to provide any input information about the business logic. Of course, the results will bear a degree of uncertainty, whereas process mining approaches typically aim at finding exact models. Still, sequence clustering can provide valuable insight into the kind of sequences that are being executed.

Process Mining Approaches

In general, all process mining approaches take an event log as input and as a starting point for the discovery of underlying processes. The event log (also called process trace or audit trail) is list of records resulting from the execution of some process. For the log to be "minable", each record usually contains information about the activity that was executed, the process instance that it belongs to, and the time of execution. The requirements on the log, i.e. the kind of information it should contain, varies according to the process mining algorithm being used.

In fact, it is the choice of mining algorithms that often leads to different process mining approaches. Some of the algorithms used for process mining include:

- the α-algorithm: an algorithm that is able to re-create the Petri-net workflow from the ordering relations found in the even log. For the algorithm to work the log must contain the process instance identifier (case id) and it must be rather complete in the sense that all ordering relations should be present in the log.
- inference methods: a set of three different algorithms used to infer a finite state machine

(FSM) from an event log, where the log is regarded as a simple sequence of symbols. The three algorithms represent different levels of compromise between accuracy and robustness to noise. The MARKOV algorithm, inspired by Markov models, seems to be the most promising. The algorithm works by building up an event graph as the result of considering Markov chains with increasing order. In the last step, the graph is converted to a FSM, which represents the process that was found.

- directed acyclic graphs: an algorithm that is able to generate a dependency graph from a workflow system log. The log must contain a relatively high number of executions of the same process so that the dependency graph for that process can be completely built. Originally, the algorithm was proposed to support the adoption of workflow systems rather than actually pursuing process mining.
- inductive workflow acquisition: an approach in which the goal is to find a hidden markov model (HMM) that best represents the structure of the original process. The HMM can be found by either top-down or bottom-up refinement of an initial HMM structure; these are known as model splitting and model merging algorithms, respectively. The initial HMM structure is built directly from the log, which is regarded as a simple sequence of symbols. Reported results suggest that model splitting is faster and more accurate than model merging.
- *hierarchical clustering:* an algorithm that, given a large set of execution traces of a single process, separates them into clusters and finds the dependency graph separately for each cluster. The clusters of workflow traces are organized into a tree, hence the concept of model hierarchy. After the workflow models for the different clusters have been found, a bottom-up pass through the tree generalizes them into a single one.
- *genetic algorithm:* an algorithm in which several candidate solutions are evaluated by a fitness function that determines how consistent each solution is with the log. Every solution is represented by a causal matrix, i.e. a map of the input and output dependencies for each activity. Candidate solutions are generated by selection, crossover and mutation as in typical genetic algorithms. The search space is the set of all possible solutions with different combinations of the activities that appear in the event log. The log should contain a relatively high number of execution traces.
- *Instance graphs:* an approach that aims at portraying graphical representations of process execution, especially using Event-driven Process Chains (EPCs). For each execution trace found in the log, an instance graph is obtained for that process instance. In order to identify possible parallelism, each instance graph is constructed using the dependencies found in the entire log. Several instance graphs can then be aggregated in order to obtain the overall model for that log.

In general, as far as input data is concerned, all these algorithms require an event log that contains several, if not a very large number, of execution traces of the same process instance. (An exception is the RNET algorithm used in [8] which can receive a single trace as training input, but the results can vary widely depending on that given input sequence.) Because the log usually contains the traces of multiple instances, it is also required to have labelling field – usually called the *case id* – which specifies the process instance for every recorded event.

Another requirement on the content of the event log is that, for algorithms such as and, which rely on finding causal relations in the log, task A can be considered the cause of task B only if B follows A but A never follows B in the log. Exceptional behavior, errors or special conditions that would make A appear after B could ruin the results. These conditions are referred to as *noise*; algorithms that are able to withstand noise are said to be robust to noise. Most algorithms can become robust to noise by discarding causal relations with probability below a given threshold; this threshold is usually one of the algorithm parameters.

The problem with these requirements is that they may be difficult to apply in many potential scenarios for process mining. For example, in some applications the *case id* may be unavailable if the log is just an unclassified stream of recorded events. In other applications, it may be useful to clearly identify and distinguish normal behavior from exceptional one, without ruling out small variations simply as noise. These issues suggest that other kind of algorithms could provide valuable insight into the original behavior that produced the log. If there is no *case id* available, and there is an unpredictable amount of ad-hoc behavior, then an algorithm that allows us to sort out and understand that behavior could be the first step before actually mining those processes. Sequence clustering algorithms are a good candidate for this job.

Sequence Clustering

Sequence clustering is a collection of methods that aim at partitioning a number of sequences into meaningful clusters or groups of similar sequences. The development of such methods has been an active field of research especially in connection with challenges in bioinformatics. Here we will present the basic principles by referring to a simple sequence clustering algorithm based on first-order Markov chains.

In this algorithm, each cluster is associated with a first-order Markov chain, where the current state depends only on the previous state. The probability that an observed sequence belongs to a given cluster is in effect the probability that the observed sequence was produced by the Markov chain associated with that cluster. For a sequence $x = \{x_0, x_1, x_2, \ldots, x_{L-1}\}$ of length L this can be expressed simply as:

$$p(x \mid c_k) = p\ (x_0\ , c_k) \cdot \prod_{i=1}^{i-L-1} p(x_i \mid x_{i-1}\ , c_k)$$

where $p(x_0\ , c_k)$ is the probability of x_0 occurring as the first state in the Markov chain associated with cluster c_k and $p(x_i \mid x_{i-1}\ , c_k)$ is the transition probability of state x_{i-1} to state xi in that same Markov chain. Given the way to compute $p(x \mid c_k)$, the sequence clustering algorithm can be implemented as an extension to the well-known Expectation-Maximization (EM) algorithm. The steps are:

1. Initialize the model parameters $p(x\ \ , c\ \ )$ and $p(x_i \mid x_{i-1}\ , c_k)$ randomly, i.e. for each cluster the state transition probabilities of the associated Markov chain are initialized at random.

2. Using the current model parameters, assign each sequence to each cluster with a probability given by equation $p(x \mid c_k) = p\ (x_0\ , ck) \cdot \prod_{i=1}^{i-L-1} p(x_i \mid x_{i-1}\ , c_k)$.

3. Use the results of step 2 to re-estimate the model parameters, i.e. recalculate the state transition probabilities of each Markov chain based on the sequences that belong to that cluster.

4. Repeat steps 2 and 3 until the mixture model converges.

The algorithm must be provided with two input tables: a case table and a nested table. The case table contains one record for each sequence; it conveys the number of sequences in the input data set together with some descriptive information about each sequence. The nested table contains the steps for all sequences, where each step is numbered and labelled. The number is the order of occurrence within the sequence, and the label is a descriptive attribute that denotes the state in a Markov chain. The case and nested tables share a one-to-many relationship: each sequence in the case table is associated with several steps in the nested table. The connecting attribute, which is application-specific, serves as key in the case table and as sequence scope delimiter in the nested table.

For the sake of clarity, let us consider a simple example. Suppose the members of a given family have different ways of zapping through TV channels according to their own interests. Let us assume that each member always finds the TV switched off, and after turning it on, goes through a set of channels before turning it off again. Every time it is turned on, the TV generates a new session identifier (session id) and records both session-related information as well as the sequence of channel changes. Figure below shows the case and nested tables for this scenario. The session identifier is both the key to the case table and the sequence scope delimiter for the nested table. The case table contains descriptive, non-sequence attributes about each session, whereas the nested table contains the steps for each sequence, both numbered and labelled.

sessionid	timeofday	sessionid	channel	numchange
1234	2007-04-09 19:41:00	1234	Music	1
1235	2007-04-09 21:15:00	1234	Music	2
1236	2007-04-10 18:56:00	1234	Movies	3
1237	2007-04-10 20:38:00	1234	Movies	4
1238	2007-04-10 21:25:00	1235	News	1
1239	2007-04-11 19:06:00	1235	Sports	2
1240	2007-04-11 20:23:00	1235	News	3

Case (a) and nested (b) tables for the simple TV usage scenario

It can be seen from this simple example that the input data to be provided to the sequence clustering algorithm already has a form of *case id*, which is the session identifier. Pre-processing techniques will have to be used to assign this *case id* if it is not available in the first place.

What is interesting to note here is the kind of results that the sequence clustering algorithm is able to produce. Figure shows four of the clusters that the algorithm was able to identify from a given set of 24 sequences for the simple TV usage scenario. Each cluster has a different Markov chain that is able to generate the sequences assigned to that cluster. This effectively captures the dominant behavior of similar sequences.

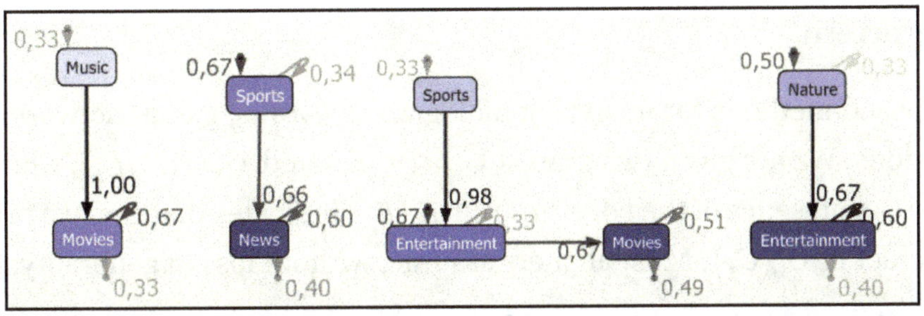

The Markov chains in four of the clusters obtained for the simple TV usage scenario.

The number of clusters to be found can be set manually or automatically by letting the algorithm perform a heuristic to determine the number of clusters for the given data. This is usually very useful to use as an initial guess before trying to run the algorithm with different parameters.

To produce the results shown in figure the algorithm performed a number of iterations, where each iteration comprises two steps: the *expectation step* and the *maximization step*. In the expectation step the algorithm assigns each sequence x to the cluster *ck* that gives the highest membership probability $p(x\,|\,c_k)$. Once this step is complete, the algorithm has a provisional estimate of which sequences belong to which cluster. In the maximization step the algorithm re-computes the transition probabilities $p(x_i\,|\,x_{i-1},c_k)$ for each cluster c_k based on the sequences that belong to that cluster. After the maximization step, the next expectation step will produce different results from the previous iteration, since $p(x\,|\,c_k)$ will now be computed with the updated values of $p(x_i\,|\,x_{i-1},c_k)$. The algorithm converges when there is no change in the values of these model parameters.

Pairwise Algorithm

Pairwise sequence alignment is a challenging task because of the exponential growth of genomic information, necessitating large-scale comparison of two input strings. The size of GenBank/the European Molecular Biology Laboratory (EMBL)/the DNA DataBank of Japan (DDBJ) nucleotide database is now doubling in every 15 months. To search databases to find out sequences similar to a given query sequence, the search programs compute an alignment score for every sequence in the database. This score represents the degree of similarity between the query and database sequence. A dynamic programming algorithm for computing the optimal local alignment score was first described by Smith and Waterman, and later improved in for linear gap penalty functions. Though dynamic programming is the best alignment procedure so far procedure so far, but it is not suitable for large strings in terms of both time and space. For two strings of lengths m and n, the time and space complexities of the Smith and Waterman (SW) algorithm are $O(\mathrm{mn})$ The time and space complexity had been improved to O (rn) in, where r is the amount of allowed error, by considering only the useful part of the distance matrix. However, for large error rates, r is $O(\mathrm{m})$, so the complexity is still $O(\mathrm{mn})$. Later on, the space complexity of SW was improved to $O(n)$. Dynamic programming has been accelerated through GLASS by first finding exactly matching long substrings, but the time and space complexity are still high. LAGAN is another implementation of dynamic programming, but is not applicable on a genome scale without prior information ("anchors") that directs comparison to orthologous regions.

Pairwise Alignment

Of the many proposed methods for analyzing biological networks, global network alignment is one of the most ambitious. We are given two graphs $G_1 = (V_1, E_1)$ and $G_2 = (V_2, E_2)$, whose vertices represent proteins, and the presence of the edge (u, v) in $E_1 (E_2)$ indicates that the two proteins represented by u and v interact in $G_1 (G_2)$. Most aligners assume, without loss of generality, that $|V_1| < |V_2|$.

The problem of pairwise alignment is to find a one-to-one function $f : V_1 \rightarrow V_2$ that maps each node in V_1 to the node in V_2 that it best matches [as a shorthand to make several equations more readable, we will treat f as a function on edges as well. In this case, $f((u,v))$ is simply more readable shorthand for $(f(u), f(v))$. It must be noted that some algorithms produce a *partial* function, abstaining from mapping nodes that cannot be matched well.

Most aligners decompose the process of producing a matching into two steps. First, for each pair of nodes in $V_1 \times V_2$, we compute their similarity by examining the local topology of the graph around those two nodes and by their sequence similarity, as measured by BLAST bit scores or E-values. Second, taking the similarities between these nodes as weighted edges in a bipartite graph with two sets of nodes V_1, V_2, we solve the maximum-weight bipartite matching problem to generate a mapping from V_1 to V_2. The pairwise alignment software available differs primarily in how they handle these two steps, with most innovation being focused on the first step of estimating the topological similarity of two nodes. This process is just a general schematic that applies to many aligners. Some, such as NATALIE 2.0, do not follow the two-step process and instead optimize a relaxed version of the problem directly.

We must stress that all existing aligners only *approximately* solve this problem, and they generally introduce approximations in two ways. First, they introduce a relaxed problem definition. Second, they use a heuristic algorithm to approximately solve the relaxed problem. For instance, GRAAL frames the problem as matching nodes to one another in such a way as to maximize their graphlet signature similarity (a measure of how many small subgraphs of various shapes are in both nodes' respective neighborhoods), in the hopes that maximizing this metric between all aligned pairs of nodes will produce a biologically informative alignment. Then, a greedy matching heuristic is used that aligns the most similar pair of nodes first and then works outward, aligning their neighborhoods. When evaluating this algorithm, it is not clear how much of its performance is attributable to the metric of graphlet signature similarity, and how much is due to using a matching algorithm that prefers to map neighbors to neighbors. All we can do is evaluate how well the heuristic solution to the relaxed problem solves our original problem. If a given aligner performs poorly, it could be that the relaxed problem is a good choice, but that choice is hamstrung by a poorly designed heuristic to solve it. Without being able to swap out the parts of these aligners, however, all we can do is evaluate their final results.

Genetic Algorithm

Genetic Algorithm (GA) is a search-based optimization technique based on the principles of Genetics and Natural Selection. It is frequently used to find optimal or near-optimal solutions to difficult problems which otherwise would take a lifetime to solve. It is frequently used to solve optimization problems, in research, and in machine learning.

Optimization

Optimization is the process of making something better. In any process, we have a set of inputs and a set of outputs as shown in the following figure.

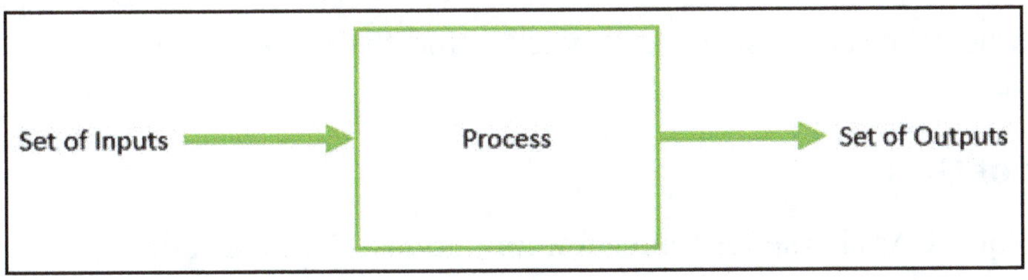

Optimization refers to finding the values of inputs in such a way that we get the "best" output values. The definition of "best" varies from problem to problem, but in mathematical terms, it refers to maximizing or minimizing one or more objective functions, by varying the input parameters.

The set of all possible solutions or values which the inputs can take make up the search space. In this search space, lies a point or a set of points which gives the optimal solution. The aim of optimization is to find that point or set of points in the search space.

Nature has always been a great source of inspiration to all mankind. Genetic Algorithms (GAs) are search based algorithms based on the concepts of natural selection and genetics. GAs are a subset of a much larger branch of computation known as Evolutionary Computation.

GAs were developed by John Holland and his students and colleagues at the University of Michigan, most notably David E. Goldberg and has since been tried on various optimization problems with a high degree of success.

In GAs, we have a pool or a population of possible solutions to the given problem. These solutions then undergo recombination and mutation (like in natural genetics), producing new children, and the process is repeated over various generations. Each individual (or candidate solution) is assigned a fitness value (based on its objective function value) and the fitter individuals are given a higher chance to mate and yield more "fitter" individuals. This is in line with the Darwinian Theory of "Survival of the Fittest".

In this way we keep "evolving" better individuals or solutions over generations, till we reach a stopping criterion.

Genetic Algorithms are sufficiently randomized in nature, but they perform much better than random local search (in which we just try various random solutions, keeping track of the best so far), as they exploit historical information as well.

Advantages of GAs

GAs has various advantages which have made them immensely popular. These include:

- Does not require any derivative information (which may not be available for many real-world problems).

- Is faster and more efficient as compared to the traditional methods.
- Has very good parallel capabilities.
- Optimizes both continuous and discrete functions and also multi-objective problems.
- Provides a list of "good" solutions and not just a single solution.
- Always gets an answer to the problem, which gets better over the time.
- Useful when the search space is very large and there are a large number of parameters involved.

Limitations of GAs

Like any technique, GAs also suffer from a few limitations. These include:

- GAs are not suited for all problems, especially problems which are simple and for which derivative information is available.
- Fitness value is calculated repeatedly which might be computationally expensive for some problems.
- Being stochastic, there are no guarantees on the optimality or the quality of the solution.
- If not implemented properly, the GA may not converge to the optimal solution.

Generic Structure of Genetic Algorithm

- Population: It is a subset of all the possible (encoded) solutions to the given problem. The population for a GA is analogous to the population for human beings except that instead of human beings, we have Candidate Solutions representing human beings.
- Chromosomes: A chromosome is one such solution to the given problem.
- Gene: A gene is one element position of a chromosome.
- Allele: It is the value a gene takes for a particular chromosome.

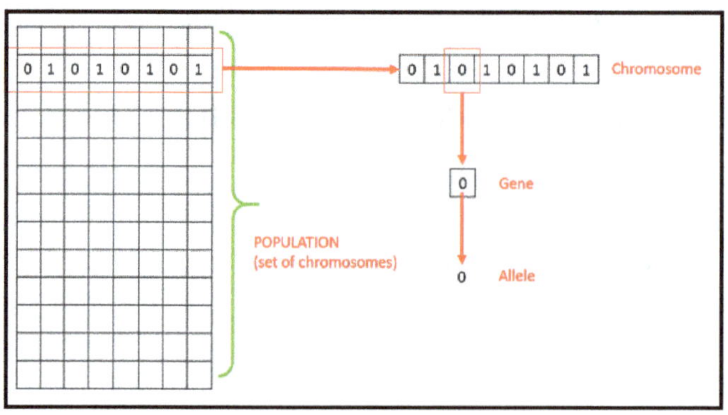

- Genotype: Genotype is the population in the computation space. In the computation space, the solutions are represented in a way which can be easily understood and manipulated using a computing system.

- Phenotype: Phenotype is the population in the actual real world solution space in which solutions are represented in a way they are represented in real world situations.
- Decoding and Encoding: For simple problems, the phenotype and genotype spaces are the same. However, in most of the cases, the phenotype and genotype spaces are different. Decoding is a process of transforming a solution from the genotype to the phenotype space, while encoding is a process of transforming from the phenotype to genotype space. Decoding should be fast as it is carried out repeatedly in a GA during the fitness value calculation.

For example, consider the 0/1 Knapsack Problem. The Phenotype space consists of solutions which just contain the item numbers of the items to be picked.

However, in the genotype space it can be represented as a binary string of length n (where n is the number of items). A 0 at position x represents that x^{th} item is picked while a 1 represents the reverse. This is a case where genotype and phenotype spaces are different.

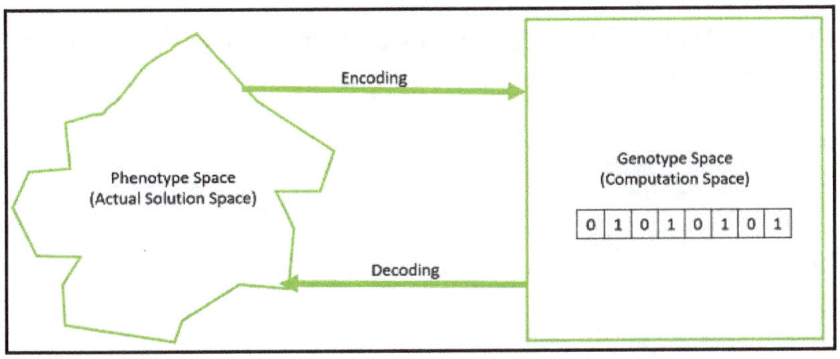

- Fitness Function: A fitness function simply defined is a function which takes the solution as input and produces the suitability of the solution as the output. In some cases, the fitness function and the objective function may be the same, while in others it might be different based on the problem.
- Genetic Operators: These alter the genetic composition of the offspring. These include crossover, mutation, selection, etc.

The basic structure of a GA is as follows:

We start with an initial population (which may be generated at random or seeded by other heuristics), select parents from this population for mating. Apply crossover and mutation operators on the parents to generate new off-springs. And finally these off-springs replace the existing individuals in the population and the process repeats. In this way genetic algorithms actually try to mimic the human evolution to some extent.

Selection

In principle, a population of individuals selected from the search space, often in a random manner, serves as candidate solutions to optimize the problem. The individuals in this population are evaluated through ("fitness") adaptation function. A selection mechanism is then used to select individuals to be used as parents to those of the next generation. These individuals will then be crossed and mutated to form the new offspring. The next generation is finally formed by an alternative mechanism between parents and their offspring. This process is repeated until a certain satisfaction condition.

Six different selection methods are considered here, namely: the roulette wheel selection (RWS), the stochastic universal sampling (SUS), the linear rank selection (LRS), the exponential rank selection (ERS), the tournament selection (TOS), and the truncation selection (TRS).

Roulette Wheel Selection (RWS)

The conspicuous characteristic of this selection method is the fact that it gives to each individual i of the current population a probability $p(i)$ of being selected, proportional to its fitness $f(i)$.

$$p(i) = \frac{f(i)}{\sum_{j=1}^{n} f(j)}$$

Where n denotes the population size in terms of the number of individuals. The RWS can be implemented according to the following pseudo-code:

RWS_ pseudo-code

{

- Calculate the sum $S = \sum_{i=1}^{n} f(j)$
- For each individual do { 1i n $\leq i \leq n$ do {

- Generate a random number $\alpha \in [0, S]$
- $i\, Sum = 0; \quad j = 0;$
- Do {
- $iSum \leftarrow iSum + f(j)$;
- $j \leftarrow +1;$

} while ($iSum < \alpha$ and $j < n$)

- Select the individual j ;}

}

Note that a well-known drawback of this technique is the risk of premature convergence of the GA to a local optimum, due to the possible presence of a dominant individual that always wins the competition and is selected as a parent.

Stochastic Universal Sampling (SUS)

The SUS is a variant of RWS aimed at reducing the risk of premature convergence. It can be implemented according to the following pseudo-code:

```
SUS_ pseudo-code
{
  • Calculate the mean f̄ = 1/n ∑_{i=1}^{n} f(j);
  • Generate a random number α ∈ [0,1];
  • Sum = f(1); delta = α × f̄; j = 0 ;
  • Do {
        • If ( delta < Sum ) {
                - select the jth individual;
                - delta = delta + Sum ;
          }
          else {
            - j = j + 1;
            - Sum = Sum + f(j) ;
          }
     } while (j < n)
}
```

Linear Rank Selection (LRS)

LRS is also a variant of RWS that tries to overcome the drawback of premature convergence of the GA to a local optimum. It is based on the rank of individuals rather than on their fitness. The rank n is accorded to the best individual whilst the worst individual gets the rank 1. Thus, based on its rank, each individual *i* has the probability of being selected given by the expression,

$$p(i) = \frac{rank(i)}{n \times (n-1)}$$

Once all individuals of the current population are ranked, the LRS procedure can be implemented according to the following pseudo-code:

```
LRS_ pseudo-code
  {
Calculate the sum v = 1/(n - 2.001);
```

```
• For each individual 1 ≤i≤n do {
    - Generate a random num α∈[0, v];
      - For each 1 ≤ j ≤ n do {
          - If ( p( j) ≤ α ) {
                - Select the j^th individual;
            - Break;
            }
        }
      }
  }
```

Exponential Rank Selection (ERS)

The ERS is based on the same principle as LRS, but it differs from LRS by the probability of selecting each individual. For ERS, this probability is given by the expression:

$$p(i)1.0*\exp\left(\frac{-rang\,(i)}{c}\right)$$

With

$$c=\frac{(n*2*(n-1))}{(6*(n-1)+n)}$$

Once the n probabilities are computed, the rest of the method can be described by the following pseudo-code:

ERS_ pseudo-code

```
{
• For each individual 1≤j≤n do {
- Generate a random number α∈[1/9 c, 2/c] ;
- For
- each 1≤j≤n do {
          - If ( p( j) ≤ α ) {
```

```
                    - Select the jth individual;
                    - Break;
                } // end if
            } // end for j
        }// end for i
}
```

Tournament Selection (TOS)

Tournament selection is a variant of rank-based selection methods. Its principle consists in randomly selecting a set of k individuals. These individuals are then ranked according to their relative fitness and the fittest individual is selected for reproduction. The whole process is repeated times for the entire population. Hence, the probability of each individual to be selected is given by the expression:

$$p(i)\begin{cases}\frac{C_{n-1}^{k-1}}{C_n^k} \text{ if } i \in [1, n-k-1] \\ 0 \text{ if } i \in [n-k, n]\end{cases}$$

Technically speaking, the implementation of TOS can be performed according to the pseudo-code:

```
TOS_ pseudo-code
{
•  Create a table t  where the n  individuals are placed in a randomly
   chosen order
•  For i = 1 to n do {
•  for j = 1 to n do {
     •  i1 = t( j) ;
     •  For m = 1 to n do {
        •  i2 = t( j + m) ;
        •  If ( f (i1 ) > f (i2 ) ) the select i1
        }// end for m
     •  j = j + k ;
     } // end for j

   }// end for i
}
```

Truncation Selection (ERS)

The truncation selection is a very simple technique that orders the candidate solutions of each population according to their fitness. Then, only a certain portion p of the fittest individuals are selected and reproduced 1/ p times. It is less used in practice than other techniques, except for very large population. The pseudo-code of the technique is as follows:

```
TRS_ pseudo-code
{
•  Order the n  individuals of fitness;
•  Set the portion p  of individuals to select (e.g.10% ≤ p ≤ 50%);
•  sp = int(n × p) // selection pressure;
•  Select the first sp  individuals;
}
```

Crossover

The crossover operator is a genetic operator that combines (mates) two chromosomes (parents) to produce a new chromosome (offspring). The idea behind crossover is that the new chromosome may be better than both of the parents if it takes the best characteristics from each of the parents. Crossover occurs during evolution according to a user definable crossover probability.

Single Point Crossover

When performing crossover, both parental chromosomes are split at a randomly determined crossover point. Subsequently, a new child genotype is created by appending the first part of the first parent with the second part of the second parent. A single crossover point on both parents' organism strings is selected. All data beyond that point in either organism string is swapped between the two parent organisms. Figure shows the single point crossover (SPC) process.

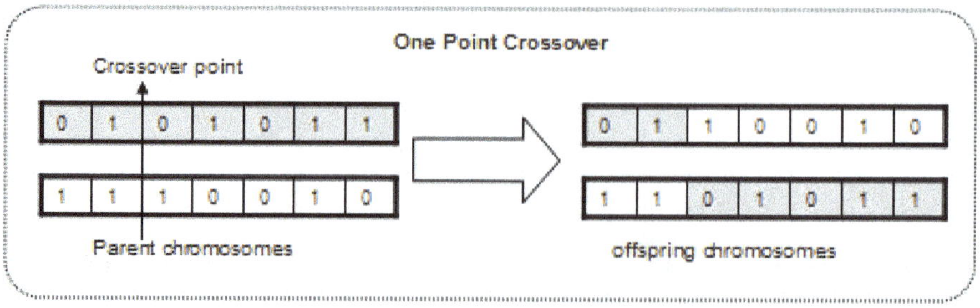

Single point crossover

Two Point Crossover

Apart from SPC, many different crossover algorithms have been devised, often involving more than one cut point. It should be noted that adding further crossover points reduces the performance of the GA. The problem with adding additional crossover points is that building blocks are more likely to be disrupted. However, an advantage of having more crossover points is that the problem space may be searched more thoroughly. In two-point crossover (TPC), two crossover points are chosen and the contents between these points are exchanged between two mated parents.

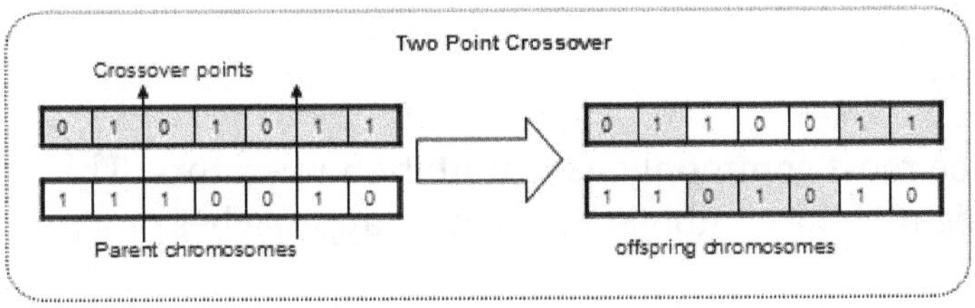

Two point crossover

In figure above, the arrows indicate the crossover points. Thus, the contents between these points are exchanged between the parents to produce new children for mating in the next generation.

Intermediate Crossover

Intermediate creates offsprings by a weighted average of the parents. Intermediate crossover (IC) is controlled by a single parameter Ratio:

$$offspring = parent1 + rand * Ratio * (parent2 - parent1)$$

If Ratio is in the range then the offsprings produced are within the hypercube defined by the parents locations at opposite vertices. Ratio can be a scalar or a vector of length number of variables. If Ratio is a scalar, then all of the offsprings will lie on the line between the parents. If Ratio is a vector then children can be any point within the hypercube.

Heuristic Crossover

In heuristic crossover (HC), heuristic returns an offspring that lies on the line containing the two parents, a small distance away from the parent with the better fitness value in the direction away from the parent with the worse fitness value. The default value of Ratio is 1.2. If parent1 and parent2 are the parents, and parent1 has the better fitness value, the function returns the child,

$$offspring = parent2 + Ratio * (parent1 - parent2)$$

Arithmetic Crossover

In arithmetic crossover (AC), arithmetic creates children that are the weighted arithmetic mean of two parents. Children are feasible with respect to linear constraints and bounds. Alpha is random value between. If parent1 and parent2 are the parents, and parent1 has the better fitness value, the function returns the child,

$$offspring = alpha * parent1 + (1 - alpha) * parent2$$

Mutation

In simple terms, mutation may be defined as a small random tweak in the chromosome, to get a new solution. It is used to maintain and introduce diversity in the genetic population and is usually applied with a low probability – p_m. If the probability is very high, the GA gets reduced to a random search.

Mutation is the part of the GA which is related to the "exploration" of the search space. It has been observed that mutation is essential to the convergence of the GA while crossover is not.

Mutation Operators

We describe some of the most commonly used mutation operators. This is not an exhaustive list and the GA designer might find a combination of these approaches or a problem-specific mutation operator more useful.

Bit Flip Mutation

In this bit flip mutation, we select one or more random bits and flip them. This is used for binary encoded GAs.

0	0	1	1	0	1	0	0	1	0	=>	0	0	1	0	0	1	0	0	1	0

Random Resetting

Random Resetting is an extension of the bit flip for the integer representation. In this, a random value from the set of permissible values is assigned to a randomly chosen gene.

Swap Mutation

In swap mutation, we select two positions on the chromosome at random, and interchange the values. This is common in permutation based encodings.

1	2	3	4	5	6	7	8	9	0	=>	1	6	3	4	5	2	7	8	9	0

Scramble Mutation

Scramble mutation is also popular with permutation representations. In this, from the entire chromosome, a subset of genes is chosen and their values are scrambled or shuffled randomly.

0	1	2	3	4	5	6	7	8	9	=>	0	1	3	6	4	2	5	7	8	9

Inversion Mutation

In inversion mutation, we select a subset of genes like in scramble mutation, but instead of shuffling the subset, we merely invert the entire string in the subset.

Schema

The theoretical background of genetic algorithms is seen most clearly using the binary string representation and the classic algorithm. First the wildcard $\star$ shall match any single component of a vector. A schema is built of the alphabet of genes (the fixed components) augmented by the wildcard symbol, e.g., $(0,1,\star)$ matches $(0,1,0)$ and $(0,1,1)$ if the alphabet is $\{0,1\}$. Hence a schema represents all strings of the search space which match the schema at all components except $\star$.

The following analysis is performed for the classic genetic algorithm, whose inner loop is the following:

1. $t := t+1$·
2. Select the new population $P(t)$ from the old one $P(t-1)$ by single string selection based on a positive fitness function.
3. Recombine $P(t)$ by one point crossover and mutation.
4. Evaluate $P(t)$.
5. Repeat.

Let $\delta(S)$ denote the defining length of a schema , i.e., the distance between the first and the last fixed component. It measures how compact the information contained in a schema is. For example $\delta\big((\star,\star,0,1,\star,1,0)\big)=4$. Let $o(S)$ denote the order of the schema S, i.e., the number of fixed components. For example $o\big((\star,\star,0,1,\star,1,0)\big)=4$. Let $\text{eval}(S,t)$ be the fitness of the schema S at time t, i.e., the average fitness of all strings in the population that are matched by S. If S matches the n individuals $\{v_1,\ldots,v_n$, then $\text{eval}(S,t) := (1/k)\sum_{k=1}^{n}\text{eval}(v_k)$, where $\text{eval}(v_k)$ is the fitness of a single individual v_k. Finally $F(t)$ is the total fitness of the whole population at time t, i.e., $F(t) := \sum_k \text{eval}(v_k)$. $N(t)$ is the size of the population at time t.

In the single string selection step of the classic genetic algorithm considered, an individual v_k has the probability $\text{eval}(v_k)/F(t)$ of being selected. After the selection step it is expected that,

$$\xi(S,t+1) = \xi(S,t)\cdot N(t)\cdot\frac{\text{eval}(S,t)}{F(t)}$$

strings match the schema S, since S matches $\xi(S,t)$ individuals, the number of single string selections is $N(t)$, and the probability that an average string matched by S is selected in a single string selection is $\text{eval}(S,t)/F(t)$.

Introducing the average fitness of the population $\bar{F}(t) := F(t)/N(t)$, this can be written as,

$$\xi(S,t+1) = \xi(S,t)\cdot\frac{\text{eval}(S,t)}{\bar{F}(t)}.$$

The interpretation of this formula is that the number of individuals matching a schema S grows in each time step like the ratio of the fitness of S and the total fitness. Thus an above average schema, i.e., a schema with $\text{eval}(S,t) > \bar{F}(t)$, will proliferate successfully in the next generation, but a below average schema will not. In the long term the formula implies that above average schemata will proliferate exponentially.

The next step after selection is recombination, i.e., crossover and mutation, where new individuals are introduced into the population. The effects of crossover are considered first. A crossover site is selected uniformly among $m-1$ possible sites, where m is the length of the vector representing an individual. Hence $\delta(S)/(m-1)$ is the probability that the schema S is destroyed if a certain vector (or chromosome) undergoes crossover, and thus,

$$p_{cs}(S) \geq 1 - p_c \frac{\delta(S)}{m-1}$$

holds for the probability $p_{cs}(S)$ of its survival, where p_c is the probability of crossover. $p_{cs}(S)$ is greater equal, and not equal, to the right hand side, since it may happen that a schema is conserved after splitting it in middle if both ancestors happen to contain parts of the schema at the right places.

Thus taking into account selection and crossover, the number of individuals matching a schema S can be estimated by,

$$\xi(S,t+1) \geq \xi(S,t) \cdot \frac{\text{eval}(S,t)}{\bar{F}(t)} \cdot \left(1 - p_c \frac{\delta(S)}{m-1}\right).$$

The final operator to be considered is mutation. Let p_m be the probability that a single component is changed. Since mutations are independent from one another, the probability that a schema S survives a mutation is,

$$p_{ms}(S) = (1-p_m)^{o(S)}.$$

Furthermore $p_{ms} \approx 1 - o(S)p_m$, since $p_m \ll 1$.

Hence the final estimation of $(S,t \quad 1)$ is,

$$\xi(S,t+1) \gtrapprox \xi(S,t) \cdot \frac{\text{eval}(S,t)}{F(t)} \cdot \left(1 - p_c \frac{(S)}{m \quad 1} - o(S)p_m\right),$$

which takes selection, crossover, and mutation in the classic genetic algorithm into account.

The Schema theorem can thus be stated as follows:

The inequality

$$\xi(S,t+1) \gtrapprox \xi(S,t) \cdot \frac{\text{eval}(S,t)}{F(t)} \cdot \left(1 - p_c \frac{\delta(S)}{m-1} - o(S)p_m\right),$$

holds with the above notation. This means that short, low order, above average schemata receive exponentially increasing trials in subsequent generations of the classic genetic algorithm and below average schemata receive exponentially decreasing trials.

Chromosome

A very common problem in adaptive control is learning to balance a pole that is hinged on a cart that can move in one dimension along a track of fixed length, as show in Figure . The control must use `bang-bang' control, that is, a force of fixed magnitude can be applied to push the cart to the left or right.

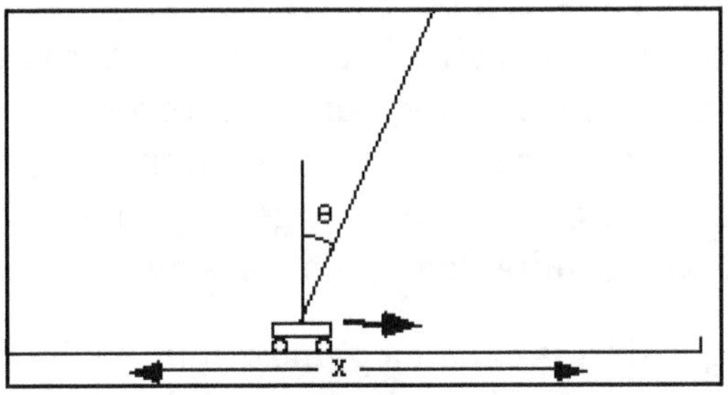

A Pole Balancer

Before we can begin to learn how to control this system, it is necessary to represent it somehow. We will use the BOXES method that was devised by Michie and Chambers (1968). The measurements taken of the physical system are the angle of the pole, θ and its angular velocity and the position of the cart, *x*, and its velocity. Rather than treat the four variables as continuous values, Michie and Chambers chose to discretize each dimension of the state space. One possible discretization is shown in figure.

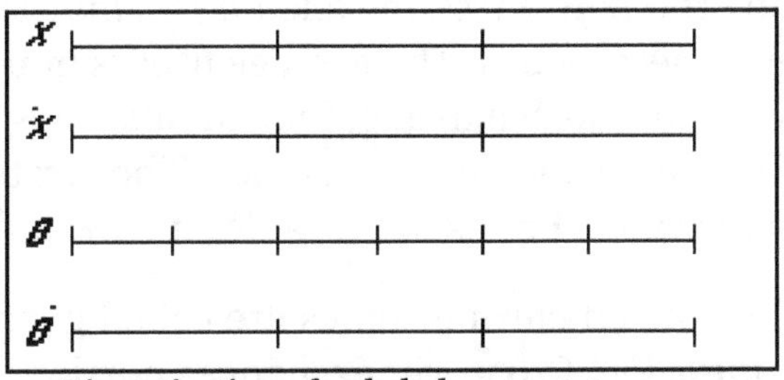

Discretization of pole balancer state space.

This discretization results in 3 x 3 x 6 x 3 = 162 `boxes' that partition the state space. Each box has associated with it an action setting which tells the controller that when the system is in that part of the state space, the controller should apply that action, which is a push to the left or a push to the right. Since there is a simple binary choice and there are 162 boxes, there are 2^{162} possible control strategies for the pole balancer.

The simplest kind of learning in this case, is to exhaustively search for the right combination. However, this is clearly impractical given the size of the search space. Instead, we can invoke a genetic search strategy that will reduce the amount of search considerably.

In genetic learning, we assume that there is a population of individuals, each one of which, represents a candidate problem solver for a given task. Like evolution, genetic algorithms test each individual from the population and only the fittest survive to reproduce for the next generation. The algorithm creates new generations until at least one individual is found that can solve the problem adequately.

Each problem solver is a *chromosome*. A position, or set of positions in a chromosome is called a *gene*. The possible values (from a fixed set of symbols) of a gene are known as *alleles*. In most genetic algorithm implementations the set of symbols is $\{0, 1\}$ and chromosome lengths are fixed. Most implementations also use fixed population sizes.

The most critical problem in applying a genetic algorithm is in finding a suitable encoding of the examples in the problem domain to a chromosome. A good choice of representation will make the search easy by limiting the search space, a poor choice will result in a large search space. For our pole balancing example, we will use a very simple encoding. A chromosome is a string of 162 boxes. Each box, or gene, can take values: 0 (meaning push left) or 1 (meaning push right). Choosing the size of the population can be tricky since a small population size provides an insufficient sample size over the space of solutions for a problem and large population requires a lot of evaluation and will be slow. In this example, 50 is a suitable population size.

Each iteration in a genetic algorithm is called a *generation*. Each chromosome in a population is used to solve a problem. Its performance is evaluated and the chromosome is given some rating of fitness. The population is also given an overall fitness rating based on the performance of its members. The fitness value indicates how close a chromosome or population is to the required solution. For pole balancing, the fitness value of a chromosome may be the number of time steps that the chromosome is able to keep the pole balanced for.

New sets of chromosomes are produced from one generation to the next. Reproduction takes place when selected chromosomes from one generation are recombined with others to form chromosomes for the next generation. The new ones are called *offspring*. Selection of chromosomes for reproduction is based on their fitness values. The average fitness of population may also be calculated at end of each generation. For pole balancing, individuals whose fitness is below average are replaced by reproduction of above average chromosomes. The strategy must be modified if two few or too many chromosomes survive. For example, at least 10% and at most 60% must survive.

Operators that recombine the selected chromosomes are called genetic operators. Two common operators are *crossover* and *mutation*. Crossover exchanges portions of a pair of chromosomes at a randomly chosen point called the crossover point. Some Implementations have more than one crossover point. For example, if there are two chromosomes, *X* and *Y*:

$$X = 1001\ 01011 \qquad Y = 1110\ 10010$$

and the crossover point is 4, the resulting offspring are:

$$O1 = 100110010 \qquad O2 = 1110\ 01011$$

Offspring produced by crossover cannot contain information that is not already in the population, so an additional operator, mutation, is required. Mutation generates an offspring by randomly changing the values of genes at one or more gene positions of a selected chromosome. For example, if the following chromosome:

$$Z = 100101011$$

is mutated at positions 2, 4 and 9, then the resulting offspring is:

$$O = 110001010$$

The number of offspring produced for each new generation depends on how members are introduced so as to maintain a fixed population size. In a *pure* replacement strategy, the whole population is replaced by a new one. In an *elitist* strategy, a proportion of the population survives to the next generation.

In pole balancing, all offspring are created by crossover (except when more the 60% will survive for more than three generations when the rate is reduced to only 0.75 being produced by crossover). Mutation is a background operator which helps to sustain exploration. Each offspring produced by crossover has a probability of 0.01 of being mutated before it enters the population. If more than 60% will survive, the mutation rate is increased to 0.25.

The number of offspring an individual can produce by crossover is proportional to its fitness:

$$\frac{\text{fitness value}}{\text{population fitness}} \times \text{No. children}$$

where the number of children is the total number of individuals to be replaced. Mates are chosen at random among the survivors.

Fitness Function

Fitness Function (also known as the Evaluation Function) evaluates how close a given solution is to the optimum solution of the desired problem. It determines how fit a solution is.

In genetic algorithms, each solution is generally represented as a string of binary numbers, known as a chromosome. We have to test these solutions and come up with the best set of solutions to solve a given problem. Each solution, therefore, needs to be awarded a score, to indicate how close it came to meeting the overall specification of the desired solution. This score is generated by applying the fitness function to the test, or results obtained from the tested solution.

Generic Requirements of a Fitness Function

The following requirements should be satisfied by any fitness function.

1. The fitness function should be clearly defined. The reader should be able to clearly understand how the fitness score is calculated.

2. The fitness function should be implemented efficiently. If the fitness function becomes the bottleneck of the algorithm, then the overall efficiency of the genetic algorithm will be reduced.

3. The fitness function should quantitatively measure how fit a given solution is in solving the problem.

4. The fitness function should generate intuitive results. The best/worst candidates should have best/worst score values.

Fitness Function for a Given Problem

Each problem has its own fitness function. The fitness function that should be used depends on the given problem. Coming up with a fitness function for the given problem is the hardest part when it comes to formulating a problem using genetic algorithms.

There is no hard and fast rule that a particular function should be used in a particular problem. However, certain functions have been adopted by data scientists regarding certain types of problems.

Typically, for classification tasks where supervised learning is used, error measures such as *Euclidean distance* and *Manhattan distance* have been widely used as the fitness function.

For optimization problems, basic functions such as sum of a set of calculated parameters related to the problem domain can be used as the fitness function.

Let's go through a few example problems and their related fitness functions.

Example — Generating Sequences

Given a set of 5 genes, which can hold one of the binary values 0 and 1, we have to come up with the sequence having all 1s. So we have to maximize the number of 1s as much as possible. This can be considered as an optimization problem. Hence, the fitness function is considered as the *number of 1s present in the genome.* If there are five 1s, then it is having maximum fitness and solves our problem. If there are no 1s, then if has the minimum fitness.The code given below shows how the fitness function is being implemented to calculate the fitness score.

```
class
Individual
{
        int fitness = 0;
        int geneLength = 5;
        int[] genes = new int[5];

        public Individual() {
          Random rn = new Random();
         //Set genes randomly for each individual
          for (int i = 0; i < geneLength; i++) {
            genes[i] = rn.nextInt() % 2;
          }
          fitness = 0;
        }
        //Calculate fitness
        public void calculateFitness() {
          fitness = 0;
          for (int i = 0; i < geneLength; i++) {
            if (genes[i] == 1) {
              ++fitness;
            }
          }
        }
      }
```

Genetic Operator

Genetic operators are used in genetic algorithms to generate diversity (*mutation*-like operators) and to combine existing solutions into others (*crossover*-like operators). The main difference among them is that the former operate on one chromosome, that is, they are *unary*, while the latter are binary operators. In GAGS, genetic operators need not be coerced into the mutation/crossover paradigm, they are just functions that act on chromosomes.

In GAGS, operators constitute a class, that is, they are not methods of the chromosome class as is usual. That means that they can be created, destroyed and changed in runtime, and besides, that they can be subclassed to create new operators. This class is designed to act as a *functor*, i.e., function-syntax objects, which have operator ( ) overloaded; this is only syntactic sugar, but allows to approach the looks of the C++ implementation to the actual algorithm.

They act in two different ways: non-directed, which means that they act on a random part of the chromosome, or directed, that is, they act on a preselected part of the chromosome. Besides, many operators consider the chromosome is divided in chunks, usually called *genes*; the size of those chunks must be pased as a parameter to their constructors.

Some operators are variable-length operators: they alter the length of the chromosome; chromosome length is always computed in every method that needs it, which means that GAGS is prepared for variable-length chromosomes. Length is always change by a discrete amount of genes. Note also that binary operators acting on variable length chromosomes can also change the length of the resulting chromosome.

An example that includes binary and unary operators is as follows

```
#include <genop.hpp>        // Chromosomes already included


main () {

  const unsigned NUMGENES = 4;

  const bitLength_t SIZEGENES = 3;

  seed_random( time( (time_t) 0 ) );// From randcl

  chrom aChromosome( NUMGENES*SIZEGENES ); // Create chromosome

  cout << "aChromosome\t\t\t" << aChromosome << endl;

  genOp creeper( SIZEGENES, genOp::CREEP ); // Apply unary genOp

  creeper( &aChromosome );

  cout << "aChromosome creeped\t\t" << aChromosome << endl;

  genOp mutator( (mutRate_t) 0.1 ); // Another unary genOp

  mutator( &aChromosome );

  cout << "aChromosome mutated 0.1\t" << aChromosome << endl;
```

```
  genOp SGA2pt( (unsigned char) 2 ); // Binary genOp, 2-point crossover
  bitString aBS ="111111111111"; //
  chrom anotherChrom( aBS ); // Define other Chromosome from string
  cout << "Crossovering with anotherChrom\n\t\t\t"
     << aChromosome << "\n\t\t\t" << anotherChrom << endl ;
  SGA2pt( &aChromosome, &anotherChrom );
  cout << "Result\t\t" << aChromosome << endl;
}
```

This program defines two unary operators: creeper, which changes a gene by plus or minus one and mutate, which does the usual thing; and a binary operator, SGA2pt, a simple two point crossover, and applies them to the chromosomes defined. A gene has been defined as a 3-bit segment, and each chromosome has got 4 genes. Output would look like this:

```
aChromosome        100011011110
aChromosome creeped   100010011110
aChromosome mutated 0.1 100010011010
Crossovering with anotherChrom
            100010011010
            111111111111
Result         100010011110
```

Al The genetic operator type is computed in two different ways: if its constructor has got unique parameters, the genetic operator type is deduced; if not, typed enums are used.

genOp::DUP	duplicates a gene with mutation
genOp::KILL	eliminates a gene
genOp::RANDINC	adds a random gene
genOp::TRANSP	transposes two genes, i.e., permutates its contents
genOp::CLONE	copies the chromosome without doint anything
genOp::MUT	usual bit-flip mutation
genOp::UXOVER	uniform crossover: interchanges bits between the two chromosomes
genOp::XOVER	usual n-point crossover: interchanges the part of the chromosome between two or more randomly generated points
genOp::GXOVER	same as before, but respects gene boundaries, which means that only whole genes are interchanged
genOp::ZAP	changes the value of a gene or part of a chromosome to another value
genOp::CREEP	changes the value of a gene by plus or minus one.

Genetic operators and modes.

Using the modes shown in table, genetic operators use three different constructors:

```
genOp( mutRate_t _rate,
   genOpMode _mode = MUT);  // Mutation and uniform crossover
genOp( bitLength_t _lenBits,
   genOpMode _mode = TRANSP); // Most operators acting on genes
genOp( unsigned char _numPts ); // number of xOver points > 1
```

Genetic operators can also be applied in a directed way, by using the applyAt method

```
const unsigned toKill = 1;
chrom thisChrom( "111000111000");
genOp killer( SIZEGENES, genOp::KILL);
killer.applyAt( toKill*SIZEGENES, &thisChrom );
cout << "Result\t\t\t" << thisChrom << endl;
```

BLAST

As gene and protein sequence databases grew at the end of the twentieth century, scientists turned to computers to help analyze this abundant and ever-growing amount of data. Today, one of the most common tools used to examine DNA and protein sequences is the Basic Local Alignment Search Tool, also known as BLAST. BLAST is a computer algorithm that is available for use online at the National Center for Biotechnology Information (NCBI) website, as well as many other sites. BLAST can rapidly align and compare a query DNA sequence with a database of sequences, which makes it a critical tool in ongoing genomic research. In recent years, the parallel development of large-scale sequencing projects and bioinformatic tools like BLAST has enabled scientists to study the genetic blueprint of life across many species, and it has also helped connect biology and computer science in the maturing field of bioinformatics.

Alignment Theory

Although the computer science principles behind BLAST have been around for some time, prior to BLAST, they had not been applied to biology. Before BLAST, alignment programs used dynamic programming algorithms, such as the Needleman-Wunsch and Smith-Waterman algorithms, that required long processing times and the use of supercomputers or parallel computer processors.

Figure depicts a Needleman-Wunsch alignment of the words "PELICAN" and "COELACANTH." The search space of the alignment is shown using a Cartesian grid and is proportional to the length of the sequences being compared plus one extra row and column.

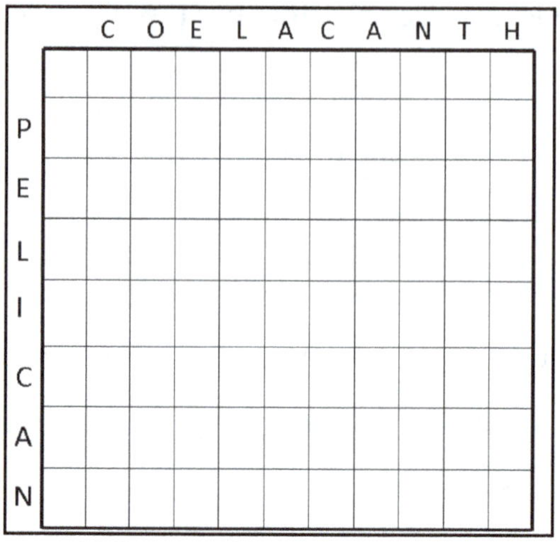

Initialization of the alignment matrix.

A Needleman-Wunsch alignment of the words “PELICAN” and “COELACANTH.”

Next, the alignment matrix is initialized with a zero in the upper left corner. For each letter of the word being aligned, a point is deducted so that each letter has a progressively more negative score. Why does the algorithm subtract a point? In an alignment, the diagonal is read from the upper left to the lower right, and when the analysis moves vertically or horizontally, it indicates a gap in the sequence. Thus, each time the program moves straight up or down, a gap penalty is applied that takes away points from the alignment score. Finally, a little arrow, or pointer, is added to indicate which direction to follow the alignment. In the third stage, the algorithm starts to actually build and score the alignment in a step called fill, or induction.

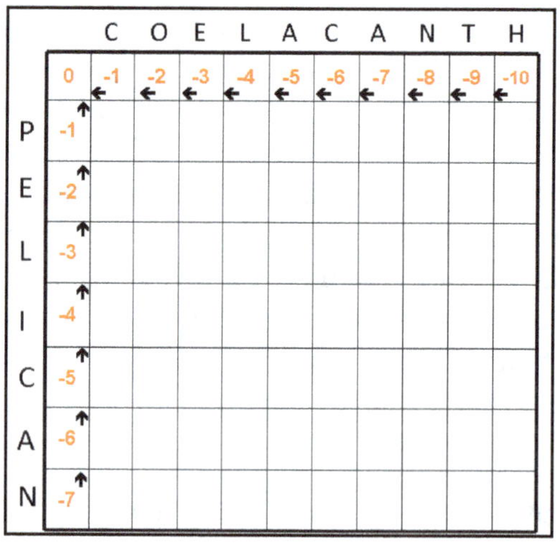

Filling the axes of the alignment matrix.

When filling the axes of the alignment matrix, start in the upper left corner and set it to 0. Next, assign a score for each letter in the row or column. Note that there is a penalty for gaps, and that the arrow should point toward the origin of the alignment.

In this example, the analysis begins by aligning the C to the P and calculating a score. In figure, one point is added if two letters match, and one point is subtracted if they do not. This calculation is carried out three times, once for each square to the left (dark blue), above (green), and to the upper left (brown). Using a value from either the upper square or the left square, the final score is -2 (-1 + -1). Using the 0 score in the upper left diagonal square, the final score is -1 (0 + -1). Because -1 is the highest score, this score is jotted down in the alignment matrix, and because the upper left square was the one leading to the best score, an arrow is inserted in the box pointing toward this square.

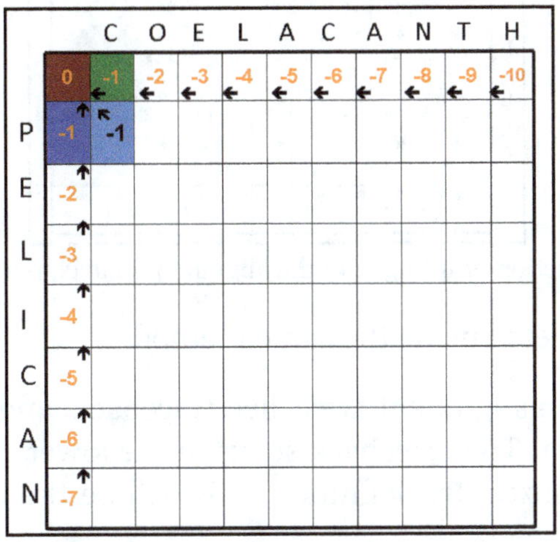

Induction or filling of the alignment matrix, part I.

One point is added if two letters match, and one point is subtracted if they do not. Using a value from either the upper (green) or left (dark blue) square, the final score is -2; however, using the value from the upper left (brown) square, the final score is -1. Because this is the highest score, it is recorded in the alignment matrix along with an arrow pointing to the upper left square.

This same process continues, calculating two scores for every square in the matrix. At the end, there is a completely scored matrix with a series of arrows used to find the optimal alignment.

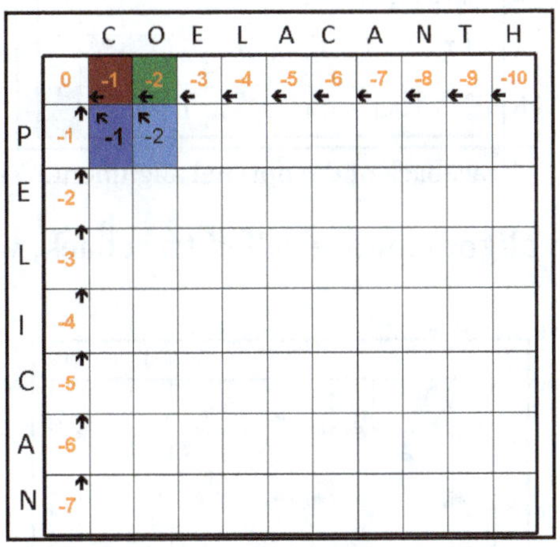

Induction or filling in of the alignment matrix, part II.

The same process is carried out for the next square in the alignment. Here, using the value in upper left (brown) square yields a sum of -2, using the value in the upper (green) square yields a sum of -3, and using the value in the left (dark blue) square yields a sum of -2. Because -2 is the highest score and was initially calculated using the upper left square, -2 is recorded in the matrix along with an arrow pointing toward the brown square.

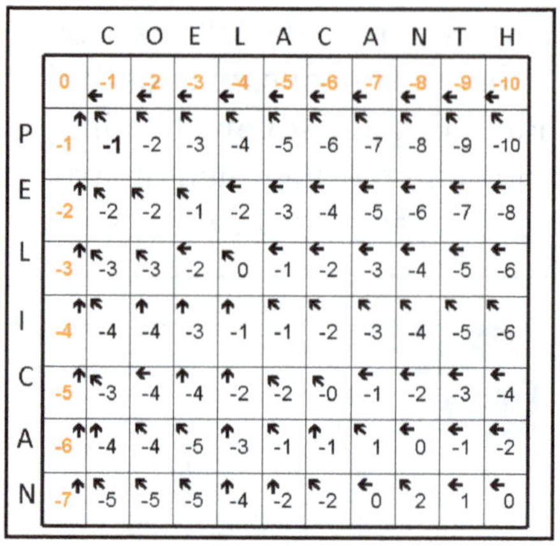

Induction or filling in of the alignment matrix, part III.

The rest of the matrix is completed using the same method.

The final steps of generating an alignment are called traceback, and they involve finding the optimal, highest-scoring alignment. The traceback starts in the lower right of the matrix and follows the pointers to the adjacent boxes. By definition, this will be the best scoring path through the alignment.

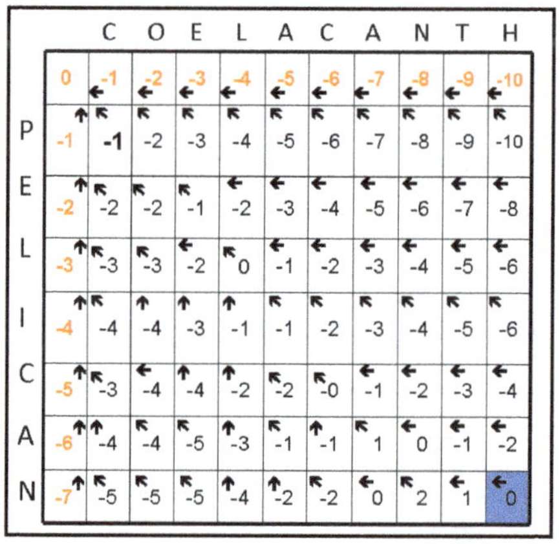

Traceback of the optimal alignment.

The final steps of generating an alignment are called traceback, which involved finding the optimal, highest-scoring alignment.

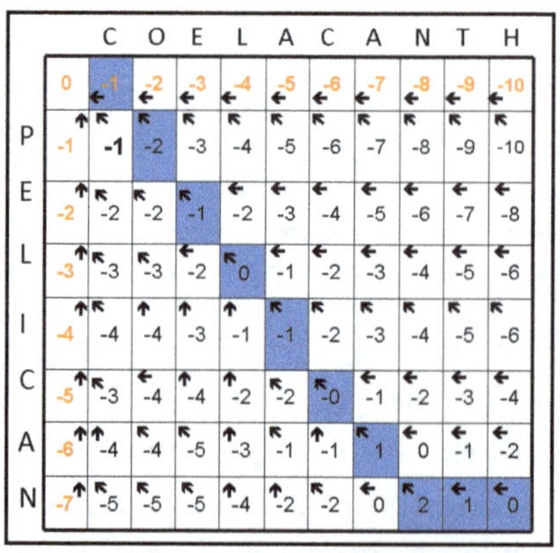

Traceb ack of the optimal complete alignment.

During traceback, following the best scoring path reveals the best alignment.

Although this sort of dynamic programming did a complete job of comparing every single residue of one sequence to every single residue of a second sequence and kept track of how well the sequences aligned at every step, these algorithms required a considerable amount of computer memory and processing time. Computing speed was an especially important concern, because these exhaustive programs had to search databases that continued to grow at exponential rates. Moreover, most regions of the search space did not score very well and therefore probably could have been skipped during the calculation process. Finally, these programs required powerful computing hardware that was expensive, rare, and ultimately impractical for most scientists and labs.

Researcher Stephen Altschul and colleagues wanted to bypass these challenges and develop a way for databases to be searched quickly on routinely used computers. In order to increase the speed of alignment, the BLAST algorithm was designed to approximate the results of an alignment algorithm created by Smith and Waterman, but to do so without comparing each residue against every other residue. BLAST is therefore heuristic in nature, meaning it has "smart shortcuts" that allow it to run more quickly. However, in this trade-off for increased speed, the accuracy of the algorithm is slightly decreased.

BLAST Heuristic

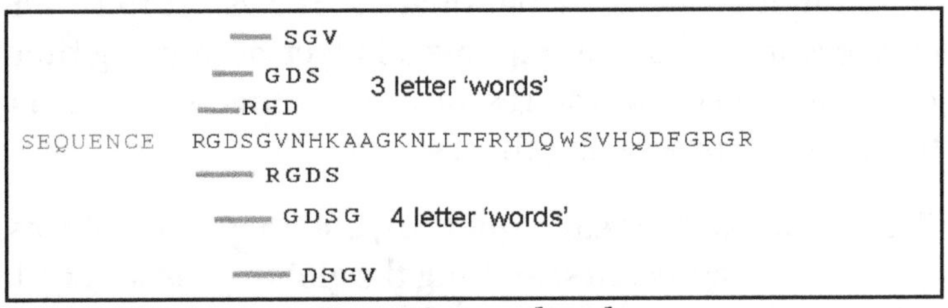

Generating word seeds.

Instead of comparing every residue against each other, BLAST uses short "word" (w) segments to create alignment "seeds." BLAST is designed to create a word list from the query sequence with words of a specific length, as defined by the user.

BLAST increases the speed of alignment by decreasing the search space or number of comparisons it makes. Specifically, instead of comparing every residue against each other, BLAST uses short "word" (w) segments to create alignment "seeds." BLAST is designed to create a word list from the query sequence with words of a specific length, as defined by the user. Requiring only three residues to match in order to seed an alignment means that fewer sequence regions need to be compared. Larger word sizes usually mean that there are even fewer regions to evaluate. Once an alignment is seeded, BLAST extends the alignment according to a threshold (T) that is set by the user. When performing a BLAST query, the computer extends words with a neighborhood score greater than T. A cutoff score (S) is used to select alignments over the cutoff, which means the sequences share significant homologies. If a hit is detected, then the algorithm checks whether w is contained within a longer aligned segment pair that has a cutoff score greater than or equal to S. When an alignment score starts to decrease past a lower threshold score (X), the alignment is terminated. These and many other variables can be adjusted to either increase the speed of the algorithm or emphasize its sensitivity.

Testing the BLAST Algorithm

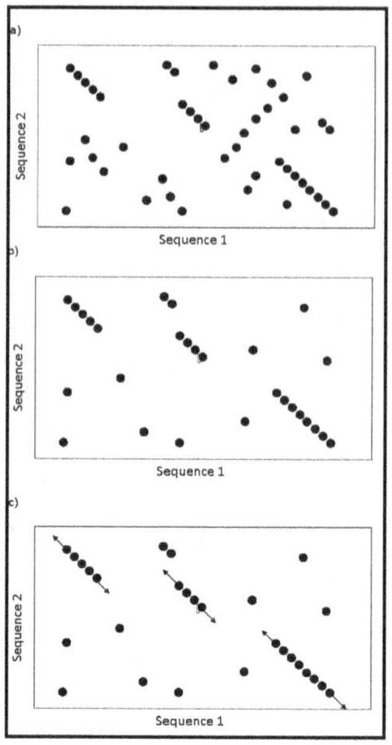

We tested the BLAST algorithm on a database of randomly generated sequences, and we examined the output resulting from different w and *T* parameters. If *T* is set to be a lower threshold, then the algorithm detects more word pairs and requires a longer processing time . Thus, choosing the value for *T*was a major decision because the researchers wanted to reach a compromise between the algorithm's sensitivity and its processing time.

Next, we tested BLAST on a database of real sequences, and they found it was successful in quickly identifying alignments with high scores. In searching the globin gene family, for example, we found that BLAST identified 88 of the 89 globin alignments that scored above 80. Other gene families, including the immunoglobulins, protein kinases, and cytochrome c genes, were then examined to measure the number of alignments detected when using different *T* and *S* values. BLAST was also able to detect similar regions within pairs of long sequences. These tests therefore showed that BLAST was fast, sensitive, and accurate as a tool for analyzing sequence alignments.

Bringing Mathematical Rigor to Alignment

One of the most notable innovations of BLAST is that the program calculates the statistical significance for each sequence alignment result. This is known as the expect value (E-value) or probability value (P-value), and it is calculated for each alignment. The E-value describes how many hits you can expect to see by chance when searching a database of a certain size, whereas the P-value describes the probability that the alignment you are observing is due to chance. In general, the lower the E- or P-value is, the more likely it is that an alignment is significant. Below the common 10^{-5} score, P and E are roughly equivalent.

The addition of statistical rigor to sequence alignment has been controversial. Some researchers rely too much on significance values to include or exclude sequences despite poorly chosen parameters, whereas others over-interpret "insignificant" results because the results "look" right. While all scientific results are subject to interpretation, BLAST scores and statistics bring much-needed objectivity to sequence comparisons, and the debate about them has helped improve methods for determining significance.

BLAST Family

Since 1990, many variants of BLAST have been developed, each with specialized features. Early on, the original BLAST was split into two adaptations: NCBI BLAST and Washington University BLAST (WU BLAST). Both BLASTs have program variations. For instance, BLASTN can be used to compare a nucleotide sequence with a nucleotide database; BLASTP can be used to compare a protein sequence with a database of protein sequences; and BLASTX can take a nucleotide sequence, translate it, and query it versus a protein database in one step. TBLASTN compares a protein query sequence to all six possible reading frames of a database and is often used to identify proteins in new, undescribed genomes. Finally, TBLASTX compares all six reading frames of a query sequence to all six reading frames of a database-an intensive algorithmic feat that can bring even modern computers to a grinding halt if not used properly.

In addition, NCBI has some of its own specialized variants of BLAST. For example, MEGABLAST is a program that can rapidly complete searches for sequences with only minor variations and can more efficiently manage queries with longer sequences. PSI- and PHI- are other powerful BLAST tools that allow more complex and evolutionary divergent proteins to be aligned.

Thus, since its creation, BLAST has become an essential tool for biologists. Its speed and sensitivity allow scientists to compare nucleotide and protein sequences to both single sequences and large databases. Most importantly, BLAST has helped democratize bioinformatics analysis and make it accessible to any researcher over the Internet; indeed, it is rare to read a modern molecular biology paper that does not refer to a BLAST alignment. In short, BLAST and its descendant applications have permitted scientists to predict the functions of genes and proteins in whole genomes, answering questions *in silico* that could never be answered at a lab bench or in the field.

BLAT

Analyzing vertebrate genomes requires rapid mRNA/DNA and cross-species protein alignments. A new tool, BLAT, is more accurate and 500 times faster than popular existing tools for mRNA/DNA alignments and 50 times faster for protein alignments at sensitivity settings typically used when comparing vertebrate sequences. BLAT's speed stems from an index of all nonoverlapping K-mers in the genome. This index fits inside the RAM of inexpensive computers, and need only be computed once for each genome assembly. BLAT has several major stages. It uses the index to find regions in the genome likely to be homologous to the query sequence. It performs an alignment between homologous regions. It stitches together these aligned regions (often exons) into larger alignments (typically genes). Finally, BLAT revisits small internal exons possibly missed at the first stage and adjusts large gap boundaries that have canonical splice sites where feasible.

BLAT, which is short for "BLAST-like alignment tool." BLAT is similar in many ways to BLAST. The program rapidly scans for relatively short matches (hits), and extends these into high-scoring pairs (HSPs). However, BLAT differs from BLAST in some significant ways. Where BLAST builds an index of the query sequence and then scans linearly through the database, BLAT builds an index of the database and then scans linearly through the query sequence. Where BLAST triggers an extension when one or two hits occur in proximity to each other, BLAT can trigger extensions on any number of perfect or near-perfect hits. Where BLAST returns each area of homology between two sequences as separate alignments, BLAT stitches them together into a larger alignment. BLAT has special code to handle introns in RNA/DNA alignments. Therefore, whereas BLAST delivers a list of exons sorted by exon size, with alignments extending slightly beyond the edge of each exon, BLAT effectively "unsplices" mRNA onto the genome—giving a single alignment that uses each base of the mRNA only once, and which correctly positions splice sites.

BLAT is used to produce the human EST and mRNA alignments. The human EST alignments compared 1.75 × 109 bases in 3.73 × 106 ESTs against 2.88 × 109 bases of human DNA and took 220 CPU hours on a Linux farm of 800 MhZ Pentium IIIs. BLAT was used in translated mode to align a 2.5× coverage unassembled whole-genome shotgun of the mouse versus the masked human genome. This involved 7.51 × 109 bases in 1.33 × 107 reads and took 16,300 CPU hours. The client/server version of BLAT is used to power untranslated and translated interactive searches on http://genome.ucsc.edu. Researchers all over the world use BLAT to perform thousands of interactive sequence searches per day. The nucleotide server has sustained over 500,000 search requests per day from program-driven queries. We do ask those researchers who are doing more than a few thousand program-driven queries to obtain a copy of BLAT to use on their own servers. The nucleotide server is not as efficient as the stand-alone program, since to save memory it does not keep the genome in memory, only the index. The index uses approximately 1 gigabyte on unmasked DNA in untranslated mode, and approximately 2.5 gigabytes on masked DNA in translated mode. The translated mode server by default is less sensitive than the default stand-alone settings. It requires three perfect amino acid 4-mers to trigger an alignment. The untranslated server usually responds to a 1000-base cDNA query in less than a second. The translated server usually responds to a 400-amino acid protein query in <5 sec.

Methods

Algorithm

All fast alignment programs break the alignment problem into two parts. Initially in a "search stage," the program detects regions of the two sequences which are likely to be homologous. The program then in an "alignment stage" examines these regions in more detail and produces alignments for the regions which are indeed homologous according to some criteria. The goal of the search stage is to detect the vast majority of homologous regions while reducing the amount of sequence that is passed to the alignment stage.

Searching with Single Perfect Matches

A simple and reasonably effective search stage is to look for subsequences of a certain size, k, which are shared by the query sequence and the database. In many practical implementations of this search, every K-mer in the query is compared against all nonoverlapping K-mers in the database. Let's examine the number of homologous regions that are missed, and the number of nonhomologous regions that are passed to the alignment stage using these criteria. First, we'll need some definitions:

K: The K-mer size. Typically this is 8–16 for nucleotide comparisons and 3–7 for amino acid comparisons.

M: The match ratio between homologous areas. This would be typically about 98% for cDNA/genomic alignments within the same species, about 89% for protein alignments between human and mouse.

H: The size of a homologous area. For a human exon this is typically 50–200 bases.

G: The size of the database—3 billion bases for the human genome.

Q: The size of the query sequence.

A: The alphabet size; 20 for amino acids, 4 for nucleotides.

Assuming that each letter is independent of the previous letter, the probability that a specific K-mer in a homologous region of the database matches perfectly the corresponding K-mer in the query is simply:

$$p_1 = M^k$$

It is convenient to introduce a term that counts the number of nonoverlapping K-mers in the homologous region:

$$\mathrm{T} = \mathrm{floor}\left(\mathrm{H}/\ \mathrm{K}\right)$$

The probability that at least one nonoverlapping K-mer in the homologous region matches perfectly with the corresponding K-mer in the query is:

$$\mathrm{P} = 1 - \left(1 - \mathrm{p}_1\right)^{\mathrm{T}} = 1 - \left(1 - \mathrm{M}^{\mathrm{K}}\right)^{\mathrm{T}}$$

The number of nonoverlapping K-mers that are expected to match by chance, assuming that all letters are equally likely to occur is:

$$\mathrm{F} = (\mathrm{Q} - \mathrm{K} + 1)^{*} (\mathrm{G}/\ \mathrm{K})^{*} (1/\ \mathrm{A})^{\mathrm{K}}$$

Tables below show P and F values for various levels of sequence identity and K-mer sizes. For EST alignments we might want the search phase to find at least 99% of sequences that have 5% or less sequencing noise. Looking at Table below, to achieve this level of sensitivity using this simple search method, we would need to choose a K of 14 or less. A K of 14 results in 399 regions passed on to the alignment phase by chance alone. Any smaller K would pass significantly more. Mouse and human sequences average 89% identity at the amino acid level. Looking at Table below to compare a translated mouse read and find at least 99% of the sequences at this level of identity we would need a K of 5 or less, which would result in 62,625 sequences passed on to the alignment stage. Depending on the cost of the alignment stage, these simple search criteria may or may not be suitable. Comparing mouse and human coding sequences at the nucleotide level, where there is on average 86% base identity, requires us to reduce our K to 7 to find at least 99% of the sequences. This results in 13,078,962 regions passed to the alignment stage, which would probably not be practical.

Sensitivity and Specificity of Single Perfect Nucleotide K-mer Matches as a Search Criterion

	7	8	9	10	11	12	13	14
A. 81%	0.974	0.915	0.833	0.726	0.607	0.486	0.373	0.314
83%	0.988	0.953	0.897	0.815	0.711	0.595	0.478	0.415
85%	0.996	0.978	0.945	0.888	0.808	0.707	0.594	0.532
87%	0.999	0.992	0.975	0.942	0.888	0.811	0.714	0.659
89%	1.000	0.998	0.991	0.976	0.946	0.897	0.824	0.782
91%	1.000	1.000	0.998	0.993	0.981	0.956	0.912	0.886
93%	1.000	1.000	1.000	0.999	0.995	0.987	0.968	0.957
95%	1.000	1.000	1.000	1.000	0.999	0.998	0.994	0.991
97%	1.000	1.000	1.000	1.000	1.000	1.000	1.000	0.999
B. K	7	8	9	10	11	12	13	14
F	1.3e+07	2.9e+06	635783	143051	32512	7451	1719	399

(*A*) Columns are for K sizes of 7–14. Rows represent various percentage identities between the homologous sequences. The table entries show the fraction of homologies detected as calculated from equation 3 assuming a homologous region of 100 bases. The larger the value of K, the fewer homologies are detected.

(*B*) K represents the size of the perfect match. F shows how many perfect matches of this size expected to occur by chance according to equation 4 in a genome of 3 billion bases using a query of 500 bases.

Sensitivity and Specificity of Single Perfect Amino Acid K-mer Matches as a Search Criterion

K	3	4	5	6	7
A. 71%	0.992	0.904	0.697	0.496	0.317
73%	0.996	0.931	0.752	0.560	0.374
75%	0.998	0.952	0.803	0.625	0.436
77%	0.999	0.969	0.850	0.689	0.503
79%	0.999	0.981	0.890	0.752	0.574
81%	1.000	0.989	0.924	0.810	0.646
83%	1.000	0.994	0.950	0.862	0.718
85%	1.000	0.997	0.970	0.906	0.787
87%	1.000	0.999	0.984	0.942	0.850
89%	1.000	1.000	0.993	0.968	0.903
91%	1.000	1.000	0.997	0.985	0.945
93%	1.000	1.000	0.999	0.995	0.975
B. K	3	4	5	6	7
F	4.2e+07	1.6e+06	62625	2609	112

(*A*) Columns are for K sizes of 3–7. Rows represent various percentage identities between the homologous sequences. The table entries show the fraction of homologies detected as calculated from equation 3 assuming a homologous region of 33 amino acids. (*B*) K represents the size of the perfect match. F shows how many perfect matches of this size are expected to occur by chance according to equation 4 in a translated genome of 3 billion bases using a query of 167 amino acids (corresponding to 500 bases).

Searching with Single almost Perfect Matches

What if instead of requiring perfect matches with a K-mer to trigger an alignment, we allow almost perfect matches, that is, hits where one letter may mismatch? The probability that a nonoverlapping K-mer in a homologous region of the database matches almost perfectly the corresponding K-mer in the query is:

$$p_1 = K^* M^{k-1} {}^*(1 - M) + M^K$$

As with a single perfect hit, the probability that any nonoverlapping K-mer in the homologous region matches almost perfectly with the corresponding K-mer in the query is:

$$P = 1 - (1 - p_1)^T$$

Whereas the number of K-mers which match almost perfectly by chance are:

$$F = (Q - K + 1)^* (G/K)^* \left(K^* (1/A)^{K-1*} (1 - (1/A)) + (1/A)^K\right)$$

Tables below show P and F for various levels of sequence identity and K-mer sizes. For the purposes of EST alignments, a K of 22 or less would pass through over 99% of the truly homologous regions while on average passing less than one chance match through to the aligner. With a reasonably fast alignment stage, it would be feasible to look for mouse/human homologies at the nucleotide level using this technique. A K size of 12 detects over 99% of the mouse homologies, and requires checking 275,671 alignments. At the amino acid level, a K size of 8 has the desired sensitivity and requires checking only 374 alignments.

Sensitivity and Specificity of Single Near-Perfect (One Mismatch Allowed) Nucleotide K-mer Matches as a Search Criterion

	12	13	14	15	16	17	18	19	20	21	22
A. 81%	0.945	0.880	0.831	0.721	0.657	0.526	0.465	0.408	0.356	0.255	0.218
83%	0.975	0.936	0.904	0.820	0.770	0.649	0.591	0.535	0.480	0.361	0.318
85%	0.991	0.971	0.954	0.900	0.865	0.767	0.719	0.669	0.619	0.490	0.445
87%	0.997	0.990	0.983	0.954	0.935	0.867	0.833	0.796	0.757	0.634	0.591
89%	1.000	0.997	0.995	0.984	0.976	0.939	0.920	0.897	0.872	0.775	0.741
91%	1.000	1.000	0.999	0.996	0.994	0.979	0.971	0.962	0.950	0.890	0.869
93%	1.000	1.000	1.000	0.999	0.999	0.996	0.994	0.991	0.988	0.963	0.954
95%	1.000	1.000	1.000	1.000	1.000	1.000	0.999	0.999	0.999	0.994	0.992
97%	1.000	1.000	1.000	1.000	1.000	1.000	1.000	1.000	1.000	1.000	1.000
B. K	12	13	14	15	16	17	18	19	20	21	22
F	275671	68775	17163	4284	1070	267	67	17	4.2	1.0	0.3

(*A*) Columns are for K sizes of 12–22. Rows represent various percentage identities between the homologous sequences. The table entries show the fraction of homologies detected as calculated by equation 6 assuming a homologous region of 100 bases. (*B*) K represents the size of the near-perfect match. F shows how many perfect matches of this size expected to occur by chance according to equation 7 in a genome of 3 billion bases using a query of 500 bases.

Sensitivity and Specificity of Single Near-Perfect (One Mismatch Allowed) Amino Acid K-mer Matches as a Search Criterion.

	4	5	6	7	8	9
A. 71%	1.000	0.992	0.946	0.823	0.725	0.515
73%	1.000	0.995	0.965	0.867	0.785	0.586
75%	1.000	0.998	0.978	0.905	0.840	0.657
77%	1.000	0.999	0.987	0.935	0.886	0.727
79%	1.000	0.999	0.993	0.959	0.924	0.791
81%	1.000	1.000	0.997	0.976	0.952	0.849
83%	1.000	1.000	0.999	0.987	0.973	0.897
85%	1.000	1.000	0.999	0.994	0.986	0.936
87%	1.000	1.000	1.000	0.997	0.994	0.964
89%	1.000	1.000	1.000	0.999	0.998	0.982
91%	1.000	1.000	1.000	1.000	0.999	0.993
93%	1.000	1.000	1.000	1.000	1.000	0.998
B. K	4	5	6	7	8	9
F	1.2E+08	6.0E+06	300078	14985	749	37

(*A*) Columns are for K sizes of 4–9. Rows represent various percentage identities between the homologous sequences. The table entries show the fraction of homologies detected. (*B*) K represents the size of the near-perfect match. F shows how many perfect matches of this size expected to occur by chance in a translated genome of 3 billion bases using a query of 167 amino acids.

Searching with Multiple Perfect Matches

Another alternative search method is to require multiple perfect matches that are constrained to be near each other. Consider a situation where the K size is 10 and there are two hits—one starting at position 10 in the query and 1010 in the database, and another starting at position 30 in the query and 1030 in the database. These two hits could easily be part of a region of homology extending from positions 10–39 in the query and 1010–1039 in the database. If we subtract the query coordinate from the database coordinate, we get a "diagonal" coordinate. Consider the search criteria that there must be N perfect matches, each no further than W letters from each other in the target coordinate, and have the same diagonal coordinate. For N = 1, the probability that a nonoverlapping K-mer in a homologous region of the database matches perfectly the corresponding K-mer in the query is simply as before:

$$p_1 = M^K$$

The probability that there are exactly n matches within the homologous region is

$$p_n = p_1^n * (1 - p_1)^{T-n} * T!/(n! * (T-n)!)$$

And the probability that there are N or more matches is the sum:

$$P = P_N + P_{N+1} + \ldots + P_T$$

The number of sets of N perfect matches that occur by chance is a little complex to calculate. For N = 1 it is easy:

$$F_1 = (Q - K + 1)^* (G/K)^* (1/A)^K$$

The probability of a second match occuring within W letters after the first is

$$S = 1 - \left(1 - (1/A)^K\right)^{W/K}$$

because the second match can occur with any of the W/K nonoverlapping K-mers in the database within W letters after the first match. We can extend this reasoning to consider the chance that the Nth match is within W letters after the $(N-1)^{th}$ match, which gives the more general relationship

$$F_N = S * F_{N-1}$$

which can be solved as

$$F_N = F_1 * S^{N-1}$$

where F_N represents the number of chance matches of N K-mers each separated by no more than W from the previous match.

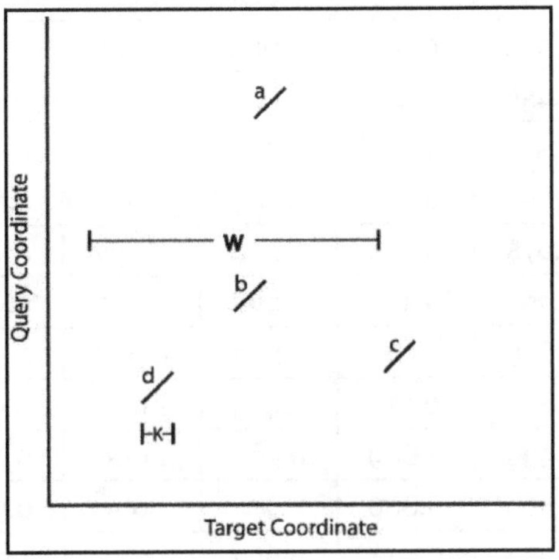

A pair of hits and two other hits. The hits a, b, c, and d are all K letters long. Hits d and b have the same diagonal coordinate and are within W letters of each other. Therefore they would match the "two perfect K-mer" search criteria.

Tables below show the sensitivity and specificity for N values of 2 and 3 and various values of other parameters which approximate cDNA or mouse/human alignments.

Sensitivity and Specificity of Multiple (2 and 3) Perfect Nucleotide K-mer Matches as a Search Criterion.

	2,8	2,9	2,10	2,11	2,12	3,8	3,9	3,10	3,11	3,12
A. 81%	0.681	0.508	0.348	0.220	0.129	0.389	0.221	0.112	0.051	0.021
83%	0.790	0.638	0.475	0.326	0.208	0.529	0.339	0.193	0.099	0.045
85%	0.879	0.762	0.615	0.460	0.318	0.676	0.487	0.313	0.180	0.093
87%	0.942	0.866	0.752	0.611	0.461	0.809	0.649	0.470	0.305	0.177
89%	0.978	0.940	0.868	0.761	0.625	0.910	0.801	0.648	0.476	0.314
91%	0.994	0.980	0.947	0.884	0.787	0.969	0.914	0.815	0.673	0.505
93%	0.999	0.996	0.986	0.962	0.912	0.993	0.976	0.933	0.851	0.722
95%	1.000	1.000	0.998	0.993	0.979	0.999	0.997	0.987	0.961	0.902
97%	1.000	1.000	1.000	1.000	0.999	1.000	1.000	0.999	0.997	0.987
B. N,K	2,8	2,9	2,10	2,11	2,12	3,8	3,9	3,10	3,11	3,12
F	524	27	1.4	0.1	0.0	0.1	0.0	0.0	0.0	0.0

(*A*) Columns are for N sizes of 2 and 3 and K sizes of 8–12. Rows represent various percentage identities between the homologous sequences. The table entries show the fraction of homologies detected as calculated by equation 10. (*B*) N and K represent the number and size of the near-perfect matches, respectively. F shows how many perfect clustered matches expected to occur by chance according to equation 14 in a translated genome of 3 billion bases using a query of 167 amino acids.

Sensitivity and Specificity of Multiple (2 and 3) Perfect Amino Acid K-mer Matches as a Search Criterion.

	2,3	2,4	2,5	2,6	2,7	3,3	3,4	3,5	3,6	3,7
A. 71%	0.945	0.643	0.297	0.126	0.044	0.945	0.643	0.297	0.126	0.044
73%	0.965	0.712	0.363	0.167	0.063	0.965	0.712	0.363	0.167	0.063
75%	0.978	0.776	0.436	0.218	0.089	0.978	0.776	0.436	0.218	0.089
77%	0.987	0.833	0.514	0.280	0.123	0.987	0.833	0.514	0.280	0.123
79%	0.993	0.882	0.596	0.353	0.169	0.993	0.882	0.596	0.353	0.169
81%	0.997	0.922	0.678	0.435	0.226	0.997	0.922	0.678	0.435	0.226
83%	0.999	0.952	0.757	0.526	0.298	0.999	0.952	0.757	0.526	0.298
85%	0.999	0.973	0.829	0.622	0.385	0.999	0.973	0.829	0.622	0.385
87%	1.000	0.987	0.889	0.719	0.485	1.000	0.987	0.889	0.719	0.485
89%	1.000	0.995	0.936	0.809	0.596	1.000	0.995	0.936	0.809	0.596
91%	1.000	0.998	0.969	0.886	0.712	1.000	0.998	0.969	0.886	0.712
93%	1.000	1.000	0.988	0.944	0.823	1.000	1.000	0.988	0.944	0.823
B. N,K	2,3	2,4	2,5	2,6	2,7	3,3	3,4	3,5	3,6	3,7
F	171875	245	0.4	0.0	0.0	708	0.0	0.0	0.0	0.0

(*A*) Columns are for N sizes of 2 and 3 and K sizes of 3–7. Rows represent various percentage identities between the homologous sequences. The table entries show the fraction of homologies detected. (*B*) N and K represents the number and size of the perfect matches, respectively. F shows how many perfect clustered matches expected to occur by chance in a translated genome of 3 billion bases using a query of 167 amino acids.

Selecting Initial Match Criteria

Both single imperfect matches and multiple perfect matches have a significant advantage over single perfect matches. They drastically reduce the number of alignments which must be checked to achieve a given level of sensitivity. The multiple-perfect match criteria can be modified to allow small insertions and deletions within the homologous area by allowing matches to be clumped if they are near each other rather than identical on the diagonal coordinate. This improves real-world sensitivity at the expense of increasing the number of alignments that must be done. Allowing a single insertion or deletion increases the alignments by a factor of three, whereas allowing two increases the alignments by a factor of five. In general, two perfect matches with the appropriate K size give specificity for a given level of sensitivity similar to that given by three or more perfect matches. The near-perfect match criterion overall is similar to the two perfect match criteria. The near-perfect criterion cannot accommodate insertions or deletions, but it has superior performance on finding small regions of homology. For finding coding exons in mouse/human alignments, whichever strategy is used, greater specificity is seen at the amino acid rather than the nucleotide level.

EST Alignment Choices

	K	F
1 Perfect	14	399
Near-Perfect	22	0.3
2 Perfect	11	0.1
3 Perfect	9	0.0

Maximum K sizes and number of chance matches passed to the alignment stage when searching for 100 bases of 95% homology with at least a 99% chance of detecting the homology. These values reflect our targets for EST alignments.

Mouse/Human Alignment Choices

	K	F	F Translated
1 Perfect-DNA	7	13,078,962	
1 Perfect-AA	5	62,625	187,875
Near-perfect-DNA	12	275,671	
Near-perfect-AA	8	749	2,247
2 Perfect DNA	6	237,983	
2 Perfect AA	4	245	734
3 Perfect DNA	5	109,707	
3 Perfect AA	3	708	2,123

Assuming 86% base identity and 89% amino acid identity, this table shows the maximum K sizes and number of chance matches passed to the alignment stage when searching for regions of 100 bases (or 33 amino acids) with at least a 99% chance of detecting the homology. These values reflect our targets for human/mouse alignments. For translated DNA sequences, the F value is multiplied by six to reflect three reading frames on both strands of the query. Even with this multiplication, the specificity for a given sensitivity is several orders of magnitude greater in the amino acid rather than the nucleotide domain.

Homology Size (In Nucleotides) vs. Sensitivity for Four Search Criteria as Applied to Mouse/Human Comparisons Assuming 86% Nucleotide Identity and 89% Amino Acid Identity.

	20	30	40	50	60	70	80	90	100	110	120	130
A	0.625	0.805	0.881	0.927	0.962	0.977	0.986	0.993	0.995	0.997	0.999	0.999
B	0.000	0.394	0.687	0.851	0.932	0.932	0.970	0.987	0.995	0.998	0.999	0.999
C	0.483	0.733	0.862	0.929	0.963	0.965	0.981	0.990	0.995	0.997	0.999	0.999
D	0.000	0.783	0.783	0.953	0.953	0.953	0.990	0.990	0.998	0.998	1.000	1.000

The median size of a human exon is 120 bases (International Human Genome Sequencing Consortium 2001). A, perfect 5 base amino acid match; B, two perfect 4-base amino acid matches within 100 amino acids and on diagonal; C, near-perfect 12 nucleotide match; D, near-perfect 8 amino acid match.

Since single-base insertions or deletions are relatively common artifacts of the sequencing process, nucleotide BLAT uses the two perfect 11-mer match criteria by default. Table shows actual alignment times for nucleotide BLAT on a collection of ESTs at various settings. For protein matches, the default criterion is a single perfect 5 for the stand-alone program. This is because the extension phase of protein BLAT is extremely quick in the stand-alone program, so the false positives generated by this approach have relatively little cost. The client/server protein BLAT uses three perfect 4-mers by default because in the client/server version, a portion of the genome must be loaded from disk for each false positive, a relatively time-consuming operation. As a result, the client/server protein BLAT is somewhat less sensitive than the stand-alone version.

Alignment Times in Seconds of 10,000 ESTs (Average Size 380 Bases) Against Human Genomic Sequence Using Various K Sizes and N Sizes.

K	N	2×10^6	2×10^7	2×10^8
10	2	3.9	35.6	680.1
10	3	3.2	21.4	348.7
11	2	2.4	8.1	92.4
11	3	2.3	6.5	61.8
12	2	3.9	7.0	39.9
12	3	3.7	6.4	33.8

The 2×10^6 genomic sequence is ctg12414, which is 2,034,363 bases long and was taken from the December 2000 UCSC human genome assembly. The 2×10^7 genomic sequence is ctg15424 and is 20,341,418 bases long. The 2×10^8 column is chromosome 4 and is 200,175,155 bases long. The two major components of the run-time are the time it takes to bin and sort the K-mer hits (clumping is almost instantaneous after sorting), and the time it takes to extend the clumps into alignments. The bin/sort time depends on the number of hits, which is proportional to 4^{-K}. The bin/sort time is somewhere between O(n) and O(n log n). The extend time is linear with respect to the number of clumps.

Clumping Hits and Identifying Homologous Regions

To implement the match criteria, BLAT builds up an index of nonoverlapping K-mers and their positions in the database. BLAT excludes K-mers that occur too often from the index, as well as K-mers containing ambiguity codes and optionally K-mers that are in lowercase rather than uppercase. BLAT then looks up each overlapping K-mer of the query sequence in the index. In this way, BLAT builds a list of "hits" where the query and the target match. Each hit contains a database position and a query position. The following algorithm is used to efficiently clump together multiple hits. The hit list is split into buckets of 64k each, based on the database position. Each bucket is sorted on the diagonal (database minus query positions). Hits that are within the gap limit are bundled together into proto-clumps. Hits within proto-clumps are then sorted along the database coordinate and put into real clumps if they are within the window limit on the database coordinate. To avoid missed clumps near the 64k bucket boundary, unclumped hits and clumps that are within the window limit are tossed into the next bucket for additional clumping opportunities. The sorting algorithm mSort, which is related to qSort, is used. The bucketing tends to keep N relatively small.

Clumps with less than the minimum number of hits are discarded, and the rest are used to define regions of the database which are homologous to the query sequence. Clumps which are within 300 bases or 100 amino acids in the database are merged together. Five hundred additional bases are added on each side to form the final homologous region.

Searching for Near Perfect Matches

BLAT has an option to allow one mismatch in a hit. This is implemented by scanning the index repeatedly for each K-mer in the query. Every possible K-mer that matches in all but one position, as well as the K-mer that matches at every position, is looked up. In all, $K*(A-1)+1$ lookups are required. For an amino-acid search with $K=8$, this amounts to 153 lookups. Because a straight index of 8-mers would require 20^8index positions or about 100 billion bytes, it is necessary to switch to a hashing scheme rather than an indexing scheme, further cutting efficiency. As a consequence, for a given level of sensitivity, the near-perfect match criterion runs 15× more slowly than the multiple-perfect match criterion in BLAT. The near-perfect match criterion seems best suited for programs that hash the query sequence rather than the database. A query sequence is sufficiently small that each possible nearly matching K-mer could be hashed, and therefore the index would not have to be scanned repeatedly.

The Relative Sensitivity and Specificity of BLAT at Various Settings With and Without a Best-Match Filter.

All nonhuman mRNA alignments		Near perfect	CPU days	% genome	% RefSeq	Enrichment
K	N					
5	1	no	8.69	0.90%	69%	77x
4	2	no	6.06	0.83%	68%	81x
8	1	yes	95.25	0.86%	69%	80x
Only the best alignment for each mRNA		Near perfect	CPU days	% Genome	% RefSeq	Enrichment
K	N					
5	1	no	8.69	0.70%	67%	95x
4	2	no	6.06	0.66%	65%	99x
8	I	yes	95.25	0.68%	66%	97x

The human genome (August 2001 freeze UCSC assembly) was aligned against a collection of 86,000 nonhuman mRNA sequences totaling 123,000,000 bases taken from Genbank. The CPU days were measured on 800 MhZ Pentiums. The percentage genome column shows what percent of bases in the human genome are part of gapless alignments with the mRNAs. The percentage RefSeq shows the percentage of coding bases in human RefSeq-defined genes (from the database at) which are part of a gapless alignment. Enrichment is the ratio of the RefSeq coding density inside of the alignments compared to coding density in the genome as a whole. Higher levels of enrichment indicate greater specificity. Also shown are the same statistics when only the single best place that a nonhuman mRNA aligns was considered.

Alignment Stage

The alignment stage performs a detailed alignment between the query sequence and the homologous regions. For historical reasons, the alignment stage for nucleotide and protein alignments is quite different. Both have limitations, and are good candidates for future BLAT upgrades. On the other hand, both are quite useful in their present form for sequences which are not too divergent.

Nucleotide Alignments

The nucleotide alignment stage is based on a cDNA alignment program first used in the Intronerator. The algorithm starts by generating a hit list between the query and the homologous region of the database. Because the homologous region is much smaller than the database as a whole, the algorithm looks for relatively small, perfect hits. If a K-mer in the query matches multiple K-mers in the region of homology, the K-mer is extended by one repeatedly until the match is unique or the K-mer exceeds a certain size. The hits are then extended as far as possible allowing no mismatches, and overlapping hits are merged. The extended hits that follow each other in both query and database coordinates are then linked together into an alignment. If there are gaps in the alignment on both the query and database side, the algorithm recurses to fill in these gaps. Because the gaps are smaller than the original query and database sequences, a smaller k can be used in generating the hit list. This continues until either the recursion finds no additional hits, or the gap is five bases or less. At this point, extensions through Ns, extensions that allow one or two mismatches if followed by multiple matches, and finally extensions that allow one or two insertions or deletions (indels) followed by multiple matches are pursued. For mRNA alignments, it is often the case that there are several equivalent-scoring placements for a large gap in the query sequence. Generally such gaps correspond to an intron. Such gaps are slid around to find their best match to the GT/AG consensus sequence for intron ends.

The nucleotide alignment strategy works well for mRNA alignments and the type of alignments needed for genomic assembly. In these cases, the sequence identity is typically 95% or better. The strategy starts to break down when base identity is below 90%, and is therefore not suitable for most cross-species alignments.

Protein Alignments

The protein alignment strategy is simpler. The hits from the search stage are kept and extended into maximally scoring ungapped alignments (HSPs) using a score function where a match is worth 2 and a mismatch costs 1. A graph is built with HSPs as nodes. If HSP A starts before HSP B in both query and database coordinates, an edge is placed from A to B. The edge is weighted by the score of B minus a gap penalty based on the distance between A and B. In the case where A and B overlap, a "crossover" point is selected which maximizes the sum of the scores of A up to the crossover and B starting at the crossover, and the difference between the full scores and the scores just up to the crossover is subtracted from the edge score. A dynamic program then extracts the maximal-scoring alignment by traversing this graph. The HSPs in the maximal-scoring alignment are removed, and if any HSPs are left the dynamic program is run again.

The major limitation of this protein alignment strategy is that if there is an indel, part of the alignment will be lost unless the search stage manages to find both sides of the indel. For the translated mouse versus translated human genome job, which was the major motivation for protein BLAT, this limitation is not as serious as it would be when searching for more distant homologs. Indeed in the translated mouse/translated human case, this limit on indels is actually useful in some ways as it reduces the amount of pseudogenes which are found by BLAT more than it reduces the amount of genes found. Even so, in the future we hope to replace this simplistic extension phase with a banded (only small gaps allowed) Smith-Waterman algorithm.

Stitching and Filling In

It is often the case that the alignment of a gene is scattered across multiple homologous regions found in the search phase. These alignments are stitched together using a minor variation of the algorithm used to stitch together protein HSPs. For DNA alignments at this stage, the gap penalty is equal to a constant plus the log of the size of the gap. For mRNA/genomic alignments, if after stitching there are gaps left between aligning blocks in both the database and query sequence, the nucleotide alignment algorithm is called on the gap to attempt to fill it in. This gives BLAT a chance to find small internal exons that are further away than 500 bases from other exons, and which are too small to be found by the search stage.

Since the sort time is O(N logN), that is, proportional to N times log N, where N is the number of hits to be sorted, and the dynamic program time is $O(N^2)$ where N is the number of HSPs, an additional step is necessary to make BLAT efficient on longer query sequences. Untranslated nucleotide queries longer than 5000 bases and translated queries longer than 1500 bases are broken into subqueries that have approximately 250 bases of overlap. Each subquery is aligned as above, and the resulting alignments are stitched together.

References

- How-to-define-a-fitness-function-in-a-genetic-algorithm-be572: towardsdatascience.com, Retrieved 17 April 2020
- Hmm-baum-welch-derivation: people.eecs.berkeley.edu, Retrieved 10 April 2020
- Selection-Methods-for-Genetic-Algorithms-259461147: researchgate.net, Retrieved 08 May 2020
- Genetic-algorithms-mutation: tutorialspoint.com, Retrieved 23 June 2020
- Basic-local-alignment-search-tool-blast-29096: nature.com, Retrieved 26 May 2020

Chapter 5

Bioanalytical Methods and Techniques

Bioanalytical techniques can be defined as the various methods used for analyzing biotics and biological matrices of biological molecules. Computational biology is one such method which employs algorithms and models to interpret biological data. Bioanalysis and bioassays are other aspects which fall under this domain. This chapter aims to provide a comprehensive understanding of all the different bioanalytical methods and techniques.

Bioanalytical Techniques

Bioanalytical techniques, as the name suggests, are the analytical tools to study the biological molecules; non-biological molecules involved with life, such as drugs; and biological processes. These tools are routinely used to identify, estimate, purify, and characterize the biomolecules. Quantification of molecules in biological samples is at the heart of bioanalysis and is routinely used to diagnose various diseases and metabolic disorders. For example, estimation of thyroxine and triiodothyronine concentrations in blood provides information about the activity of thyroid gland. Home pregnancy test kits look for the human chorionic gonadotropin (hCG) hormone in the urine, presence of which above a threshold concentration is an indicator of pregnancy. Bioanalytical methods are also used to detect drugs and their metabolites in biological samples. Initially, nonspecific assays were used to quantify the drugs in biological samples. Evolution of the existing assays, advancement in instrumentation, and introduction of newer techniques have made it possible to distinguish the drug molecules and their closely related metabolites in complex biological specimens.

Estimation of the Analytes

Identification and quantification of analytes is perhaps the most common application of bioanalytical methods. Various diseases and disorders including cancers are diagnosed by estimating the levels of the characteristic biomarkers in a particular tissue or organ. Semenogelase, for example, is a biomarker for prostate cancer, one of the most frequently diagnosed cancers in human males.

Qualitative Versus Quantitative Analyses

A qualitative analysis simply tells about the presence or absence of an analyte in a sample. An absence of analyte, however, may result due to concentrations below the detection level of the bioanalytical technique used. Qualitative analyses are used wherein detection of an analyte is sufficient to take further course of action. For example, identification of a banned performance-enhancing drug in athletics is sufficient enough to determine the qualification of the athlete to participate in the event. In certain cases, however, it is important to estimate the concentration of the analyte. A quantitative analysis would result in the determination of actual amount of the substance present in the sample. Consider a person suspected to be diabetic. A qualitative test for glucose is not good enough to ascertain if the person is diabetic. It is important to accurately determine the concentration of glucose in the blood to arrive at a conclusion. Breath alcohol detectors are used by traffic personnel to quantify the breath alcohol level, which in turn is proportional to blood glucose level and thereby enable them to identify the drunk drivers.

Accurate and Precise Determination of Analytes

It is hardly necessary to explain how critical an accurate determination of an analyte is. If a breath alcohol detector is not accurate, a drunk driver may be let off risking the life of others while a sober one may be detained. Unless the concentration of analyte is determined accurately and precisely, it is difficult to make meaningful conclusions. So, what exactly do the accuracy and precision mean? Accuracy is the measure of how closely the measured values match the true values. Precision tells about the reproducibility of the measurement *i.e.* how closely the measured values are if repeated measurements are made on the sample figure given below.

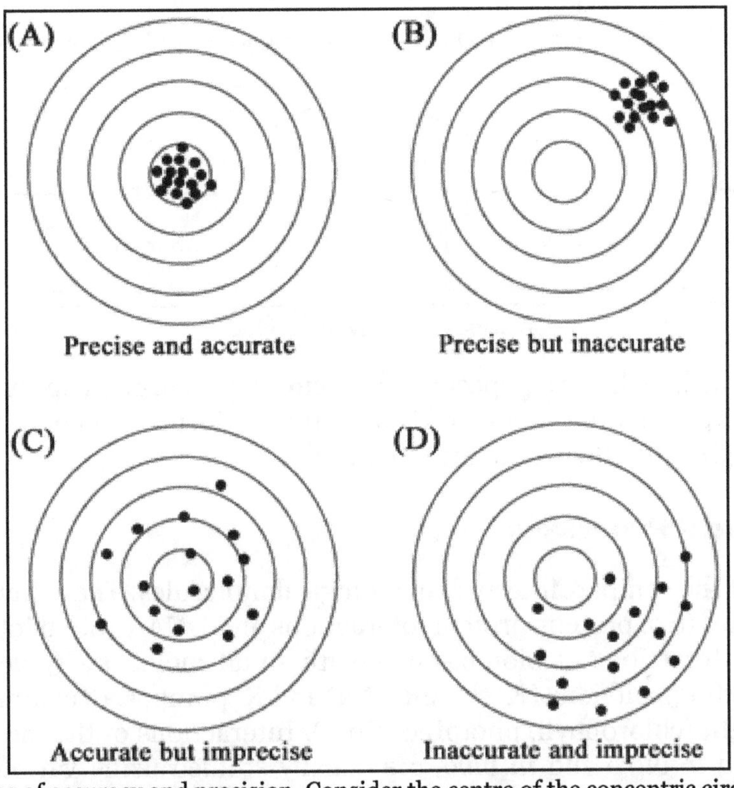

Schematic representations of accuracy and precision. Consider the centre of the concentric circles as the true value; the measured values are represented as the black dots. The measured values shown in panel A are close to the true value (accurate) as well as to each other (precise). The measured values in panel B are close to each other (precise) but far from the true value (inaccurate). The individual values in panel C are far away from the true value but randomly distributed about the true value; the average value lies close to the true value (accurate but imprecise). Panel D represents inaccurate and imprecise measurements.

It is easy to imagine the consequences of using an inaccurate equipment; it would give inaccurate results. Imprecise equipments, even if accurate, are problematic as a large number of measurements are required to arrive close to the true value which may take considerable amount of time. An analytical tool therefore has to be both accurate and precise to be used reliably and for faster analysis.

Identification and Characterization of Molecules

Researchers involved in the discovery of novel bioactive natural products often have to identify the bioactive component present in the crude sample; for example, isolation of novel antibiotics and antimicrobial peptides from various organisms. Individual components in the crude sample are isolated based on the differences in their physical and chemical properties. The bioactive component is identified by testing the activities of these isolated compounds. The bioactive compound is then characterized using various spectroscopic methods to arrive at its structure and function(s). Bioanalytical techniques can typically be classified as shown in the figure below.

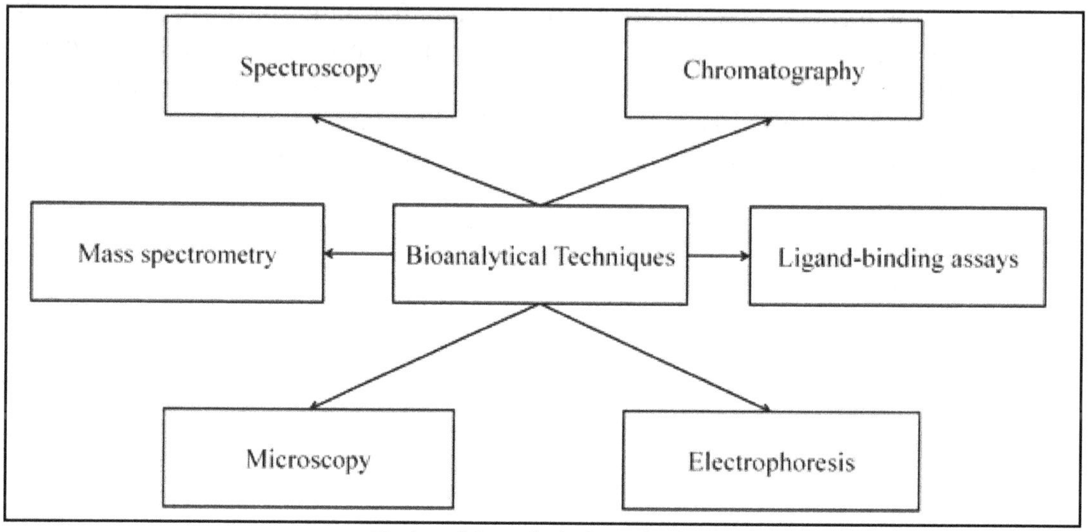

Various bioanalytical methods

Spectroscopic tools such as infrared spectroscopy, circular dichroism spectroscopy, and nuclear magnetic resonance spectroscopy can provide structural information about the molecules which in turn provides insights into their functional aspects.

Studying Biological Processes

Life is an outcome of the complex interplay of biological molecules. These involve interactions between macromolecules (*e.g.* protein-protein interactions and DNA-protein interactions, RNA-protein interactions); interactions of biomolecules with small molecules (glucose channels, water channels, ligand-binding) and ions (K^+ channel, Na^+ and K^+ pump, Ca^{2+} channels); and interaction of molecules with light (chlorophyll, photoreceptors). Interactions of the molecules with their receptors/ligands, both *in vitro* and *in vivo*, are usually studied using various spectroscopic and microscopic tools. Fluorescence spectroscopy and microscopy are among the most commonly employed tools to study the biological processes. Discovery of the green fluorescent protein (GFP) and subsequent development of its analogs with different spectral properties have revolutionized the area of cellular research. Before discussing in detail the various tools that have gained importance in bioanalytical research, it is worthwhile to take a pause for very quickly reviewing the important structural aspects of major classes of biomolecules.

Features of Major Biomolecules

Classification of biomolecules is largely based on their chemistry. There are four major classes of biomolecules: proteins, nucleic acids, carbohydrates, and lipids.

Amino Acids and Proteins

Proteins constitute the functional machinery in the living systems by carrying out most of the biological reactions. They are the unbranched polymers of L-α-amino acids. D-amino acids do exist in nature, but such molecules are rare. The structure of a typical amino acid is shown in (Figure A).

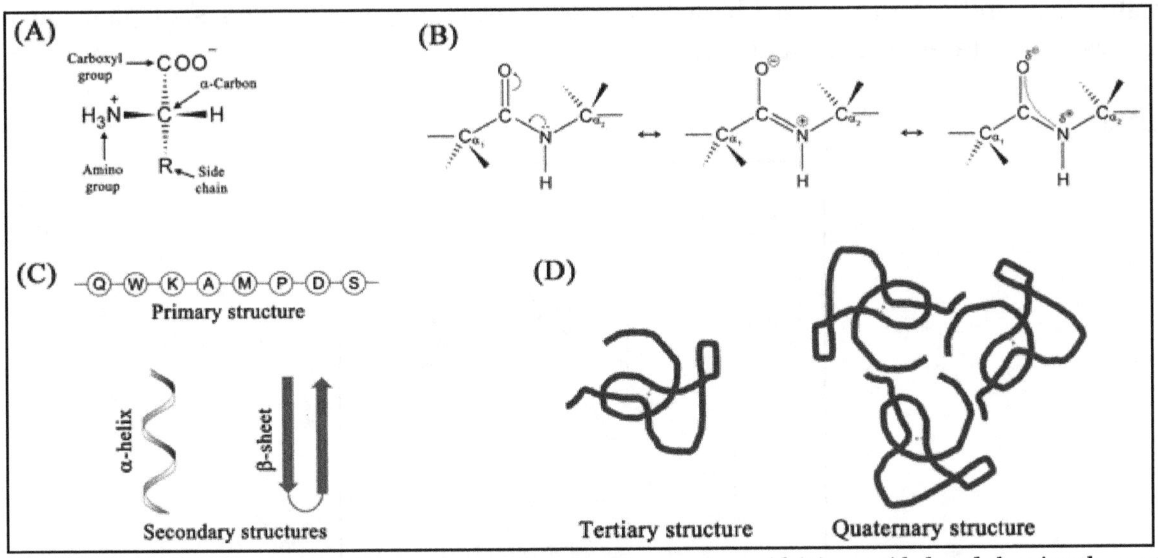

Structures of amino acids and proteins: structure of a typical L-α-amino acid (A); peptide bond showing the partial double bond character (B); primary and secondary structures (C); and tertiary and quaternary structures (D) formed by proteins.

The R group (shown in Figure A) is what differentiates the 20 standard amino acids present in proteins. During protein synthesis, the amino acids are linked together through an amide bond, called peptide bond (Figure B). Delocalization of nitrogen's lone pair of electrons over carbonyl group imparts a partial double bond character to the peptide bond putting severe conformational constraints on the polypeptide backbone (Figure B). The sequence of amino acids in a polypeptide chain is termed as its primary structure (Figure C). The linear polypeptide chain can adopt local higher order structures stabilized through hydrogen bonds; these local ordered structures are termed secondary structures (Figure C). Two such secondary structures found in proteins are: α-helices and β-sheets. Further folding of the unstructured regions in the polypeptide chain results in a compact structure, termed the tertiary structure, the highest structural level of a single chain protein (Figure D). Certain proteins function as multimers wherein more than one polypeptide chains assemble together through non-covalent interactions to form what is called a quaternary structure (Figure D).

Biomarker

A biomarker, or biological marker, generally refers to a measurable indicator of some biological state or condition. The term is also occasionally used to refer to a substance the presence of which indicates the existence of a living organism. Further, life forms are known to shed unique chemicals, including DNA, into the environment as evidence of their presence in a particular location.

Biomarkers are often measured and evaluated to examine normal biological processes, pathogenic processes, or pharmacologic responses to a therapeutic intervention. Biomarkers are used in many scientific fields.

History

The widespread use of the term "biomarker" dates back to as early as 1980. The term "biological marker" was introduced in 1950s. In 1998, the National Institutes of Health Biomarkers Definitions Working Group defined a biomarker as "a characteristic that is objectively measured and evaluated as an indicator of normal biological processes, pathogenic processes, or pharmacologic responses to a therapeutic intervention."

Medicine

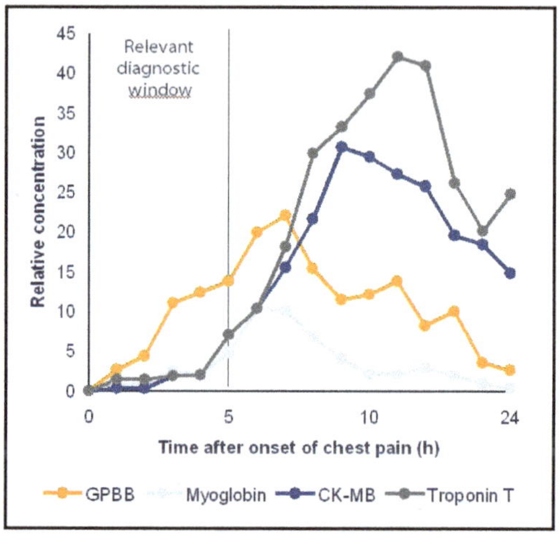

After a heart attack a number of different cardiac biomarkers can be measured to determine exactly when an attack occurred and how severe it was.

In medicine, a biomarker can be a traceable substance that is introduced into an organism as a means to examine organ function or other aspects of health. For example, rubidium chloride is used as a radioactive isotope to evaluate perfusion of heart muscle. It can also be a substance whose detection indicates a particular disease state, for example, the presence of an antibody may indicate an infection. More specifically, a biomarker indicates a change in expression or state of a protein that correlates with the risk or progression of a disease, or with the susceptibility of the disease to a given treatment.

Biochemical biomarkers are often used in clinical trials, where they are derived from bodily fluids that are easily available to the early phase researchers. A useful way of finding genetic causes of diseases such as schizophrenia has been the use of a special kind of biomarker called an endophenotype.

Other biomarkers can be based on measures of the electrical activity of the brain (using Electroencephalography (so-called Quantitative electroencephalography (qEEG)) or Magnetoencephalography), or volumetric measures of certain brain regions (using Magnetic resonance imaging) or saliva testing of natural metabolites, such as saliva nitrite, a surrogate marker for nitric oxide. One example of a commonly used biomarker in medicine is prostate-specific antigen (PSA). This marker can be measured as a proxy of prostate size with rapid changes potentially indicating cancer. The most extreme case would be to detect mutant proteins as cancer specific biomarkers through Selected Reaction Monitoring (SRM), since mutant proteins can only come from an existing tumor, thus providing ultimately the best specificity for medical purposes.

Biomarkers used for personalized medicine are typically categorized as either prognostic or predictive. An example is KRAS, an oncogene that encodes a GTPase involved in several signal transduction pathways. Prognostic biomarkers indicate the likelihood of patient outcome regardless of a specific treatment. Predictive biomarkers are used to help optimize ideal treatments, and indicates the likelihood of benefiting from a specific therapy. Biomarkers for precision oncology are typically utilized in the molecular diagnostics of chronic myeloid leukemia, colon, breast, and lung cancer, and in melanoma.

Biomarker Regulatory Validation

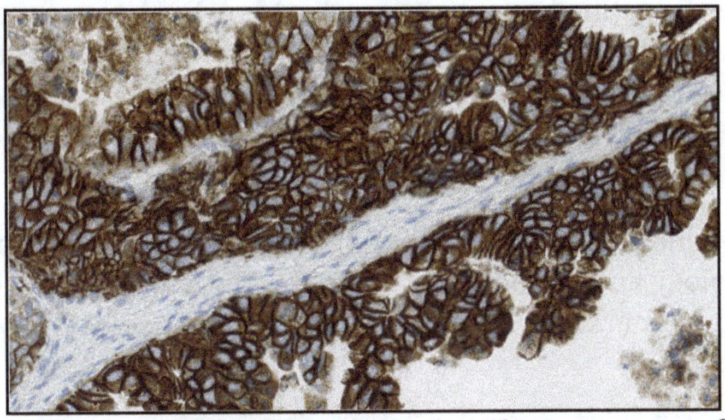

HER2 is the most popular validated biomarker. The measurement of HER2 status is used to predict breast cancer treatment responses

Proof of Concept

Previously used to identify the specific characteristics of the biomarker, this step is essential for doing an *in situ* validation of these benefits. A large number of candidates must be tested to select the most relevant ones.

Experimental Validation

This step allows the development of the most adapted protocol for routine use of the biomarker. Simultaneously, it is possible to confirm the relevance of the protocol with various methods (histology, PCR, ELISA etc.) and to define strata based on the results.

Analytical Performances Validation

One of the most important steps, it serves to identify specific characteristics of the candidate biomarker before developing a routine test. Several parameters are considered including:

- Sensitivity,
- Specificity,
- Robustness,
- Accuracy,
- Reproducibility.

Protocol Standardization

This optimizes the validated protocol for routine use, including analysis of the critical points by scanning the entire procedure to identify and control the potential risks.

Cell Biology

In cell biology, a biomarker is a molecule that allows the detection and isolation of a particular cell type (for example, the protein Oct-4 is used as a biomarker to identify embryonic stem cells).

In genetics, a biomarker (identified as genetic marker) is a DNA sequence that causes disease or is associated with susceptibility to disease. They can be used to create genetic maps of whatever organism is being studied.

Geology and Astrobiology

A biomarker can be any kind of molecule indicating the existence, past or present, of living organisms. In the fields of geology and astrobiology, biomarkers, versus geomarkers, are also known as biosignatures. The term biomarker is also used to describe biological involvement in the generation of petroleum.

Ecotoxicology

Biomarkers are used to indicate an exposure to or the effect of xenobiotics which are present in the environment and in organisms. The biomarker may be an external substance itself (e.g. asbestos particles or NNK from tobacco), or a variant of the external substance processed by the body (a metabolite) that usually can be quantified.

Nucleotide

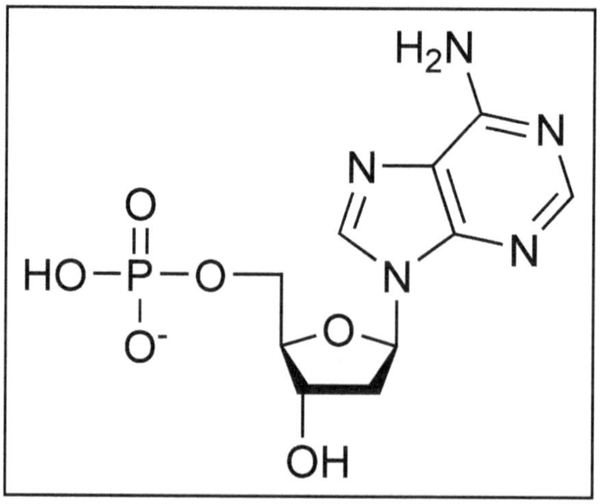

This nucleotide contains the five-carbon sugar deoxyribose (at center), a nitrogenous base called adenine (upper right), and one phosphate group (left). The sugar joined with the base forms a nucleoside called deoxyadenosine. The whole structure is a nucleotide constituent of DNA, with the name deoxyadenosine monophosphate.

Nucleotides are organic molecules that serve as the monomer units for forming the nucleic acid polymers DNA (deoxyribonucleic acid) and RNA (ribonucleic acid), both of which are essential biomolecules in all life-forms on Earth. Nucleotides are the building blocks of nucleic acids; they are composed of three subunit molecules: a nitrogenous base, a five-carbon sugar (ribose or deoxyribose), and at least one phosphate group. They are also known as *phosphate* nucleotides.

A nucleoside is a nitrogenuous base and a 5-carbon sugar. Thus a nucleoside plus a phosphate group yields a nucleotide.

Nucleotides also play a central role in life-form metabolism at the fundamental, cellular level. They carry packets of chemical energy—in the form of the nucleoside triphosphates ATP, GTP, CTP and UTP—throughout the cell to the many cellular functions that demand energy, which include synthesizing amino acids, proteins and cell membranes and parts; moving the cell and moving cell parts, both internally and intercellularly; dividing the cell, etc. In addition, nucleotides participate in cell signaling (cGMP and cAMP), and are incorporated into important cofactors of enzymatic reactions (e.g. coenzyme A, FAD, FMN, NAD, and $NADP^+$).

In experimental biochemistry, nucleotides can be radiolabeled with radionuclides to yield radionucleotides.

Structure

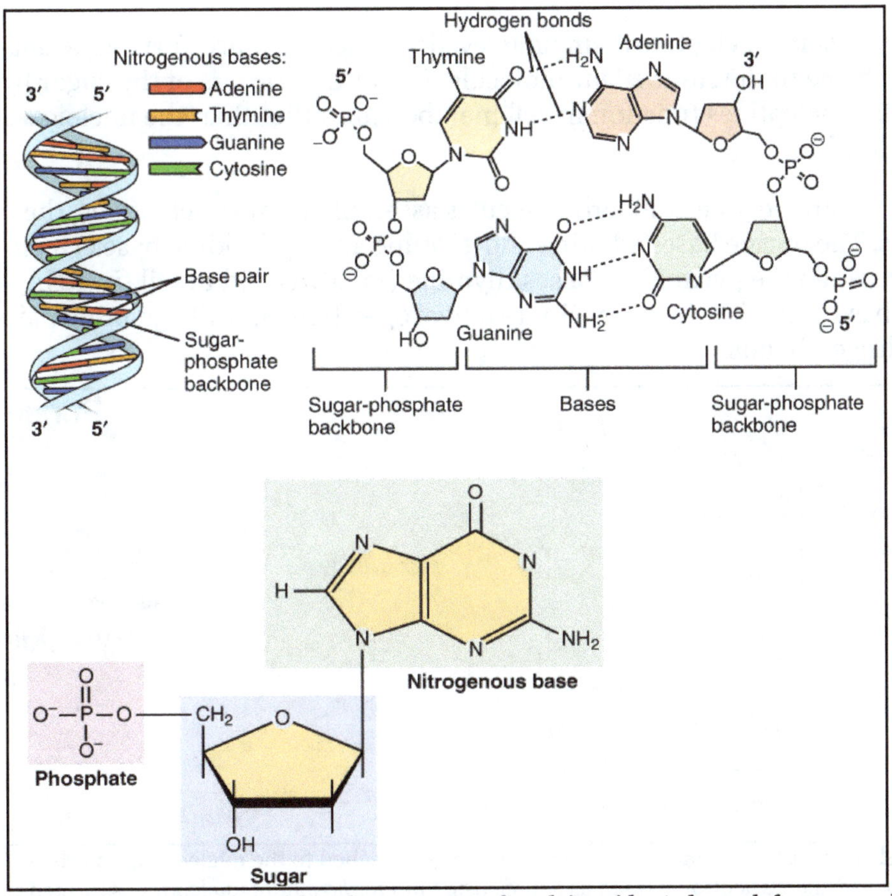

Showing the arrangement of nucleotides within the structure of nucleic acids: At lower left, a monophosphate nucleotide; its nitrogenous base represents one side of a base-pair. At upper right, four nucleotides form two base-pairs: thymine and adenine (connected by *double* hydrogen bonds) and guanine and cytosine (connected by *triple* hydrogen bonds). The individual nucleotide monomers are chain-joined at their sugar and phosphate molecules, forming two 'backbones' (a double helix) of a nucleic acid, shown at upper left.

A nucleotide is composed of three distinctive chemical sub-units: a five-carbon sugar molecule, a nitrogenous base—which two together are called a nucleoside—and one phosphate group. With all three joined, a nucleotide is also termed a "nucleoside *mono*phosphate". The chemistry sources ACS Style Guide and IUPAC Gold Book prescribe that a nucleotide should contain only one phosphate group, but common usage in molecular biology textbooks often extends the definition to include molecules with two, or with three, phosphates. Thus, the terms "nucleoside *di*phosphate" or "nucleoside *tri*phosphate" may also indicate nucleotides.

Nucleotides contain either a purine or a pyrimidine base—i.e., the nitrogenous base molecule, also known as a nucleobase—and are termed *ribo*nucleotides if the sugar is ribose, or *deoxyribo*-nucleotides if the sugar is deoxyribose. Individual phosphate molecules repetitively connect the sugar-ring molecules in two adjacent nucleotide monomers, thereby connecting the nucleotide monomers of a nucleic acid end-to-end into a long chain. These chain-joins of sugar and phosphate molecules create a 'backbone' strand for a single- or double helix. In any one strand, the chemical orientation (directionality) of the chain-joins runs from the 5'-end to the 3'-end (*read*: 5 prime-end to 3 prime-end)—referring to the five carbon sites on sugar molecules in adjacent nucleotides. In a double helix, the two strands are oriented in opposite directions, which permits base pairing and complementarity between the base-pairs, all which is essential for replicating or transcribing the encoded information found in DNA.

Unlike in nucleic acid nucleotides, singular cyclic nucleotides are formed when the phosphate group is bound twice to the same sugar molecule, i.e., at the corners of the sugar hydroxyl groups. These individual nucleotides function in cell metabolism rather than the nucleic acid structures of long-chain molecules.

Nucleic acids then are polymeric macromolecules assembled from nucleotides, the monomer-units of nucleic acids. The purine bases adenine and guanine and pyrimidine base cytosine occur in both DNA and RNA, while the pyrimidine bases thymine (in DNA) and uracil (in RNA)in just one. Adenine forms a base pair with thymine with two hydrogen bonds, while guanine pairs with cytosine with three hydrogen bonds.

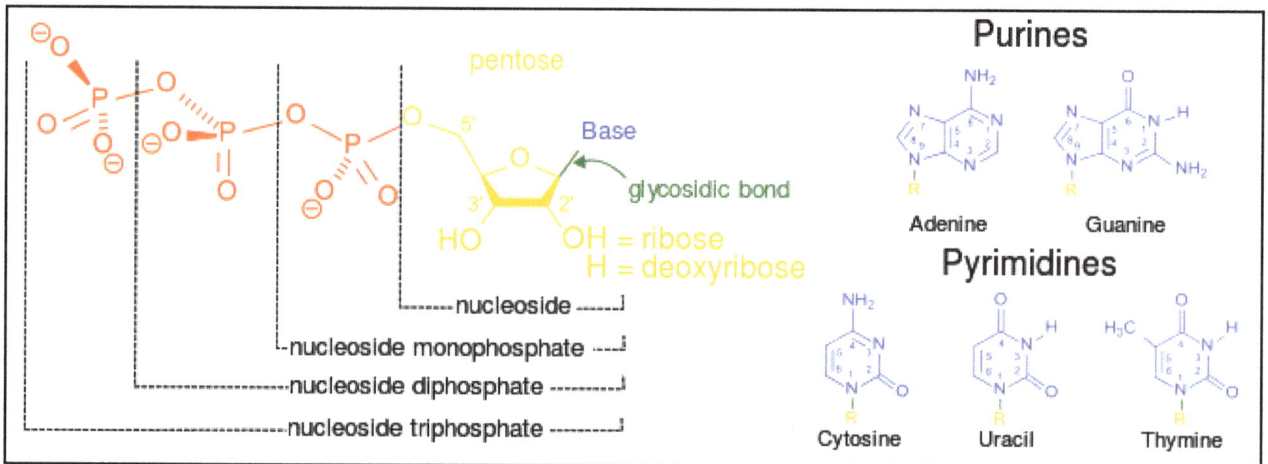

Structural elements of three nucleotides—each in turn is attached to the nucleoside (in yellow, blue, green) at center: 1st, the nucleotide termed as a *nucleoside monophosphate nucleotide* is formed by adding a phosphate group (in red); 2nd, adding a second phosphate group forms a *nucleoside diphosphate nucleotide*; 3rd, adding a third phosphate group results in a *nucleoside triphosphate nucleotide*. + The nitrogenous base (nucleobase) is indicated by "Base" and "glycosidic bond" (sugar bond). All five primary, or canonical, bases—the purines and pyrimidines—are sketched at right (in blue).

Synthesis

Nucleotides can be synthesized by a variety of means both in vitro and in vivo.

In vivo, nucleotides can be synthesized de novo or recycled through salvage pathways. The components used in de novo nucleotide synthesis are derived from biosynthetic precursors of carbohydrate and amino acid metabolism, and from ammonia and carbon dioxide. The liver is the major organ of de novo synthesis of all four nucleotides. De novo synthesis of pyrimidines and purines follows two different pathways. Pyrimidines are synthesized first from aspartate and carbamoyl-phosphate in the cytoplasm to the common precursor ring structure orotic acid, onto which a phosphorylated ribosyl unit is covalently linked. Purines, however, are first synthesized from the sugar template onto which the ring synthesis occurs. For reference, the syntheses of the purine and pyrimidine nucleotides are carried out by several enzymes in the cytoplasm of the cell, not within a specific organelle. Nucleotides undergo breakdown such that useful parts can be reused in synthesis reactions to create new nucleotides.

In vitro, protecting groups may be used during laboratory production of nucleotides. A purified nucleoside is protected to create a phosphoramidite, which can then be used to obtain analogues not found in nature and/or to synthesize an oligonucleotide.

Pyrimidine Ribonucleotide Synthesis

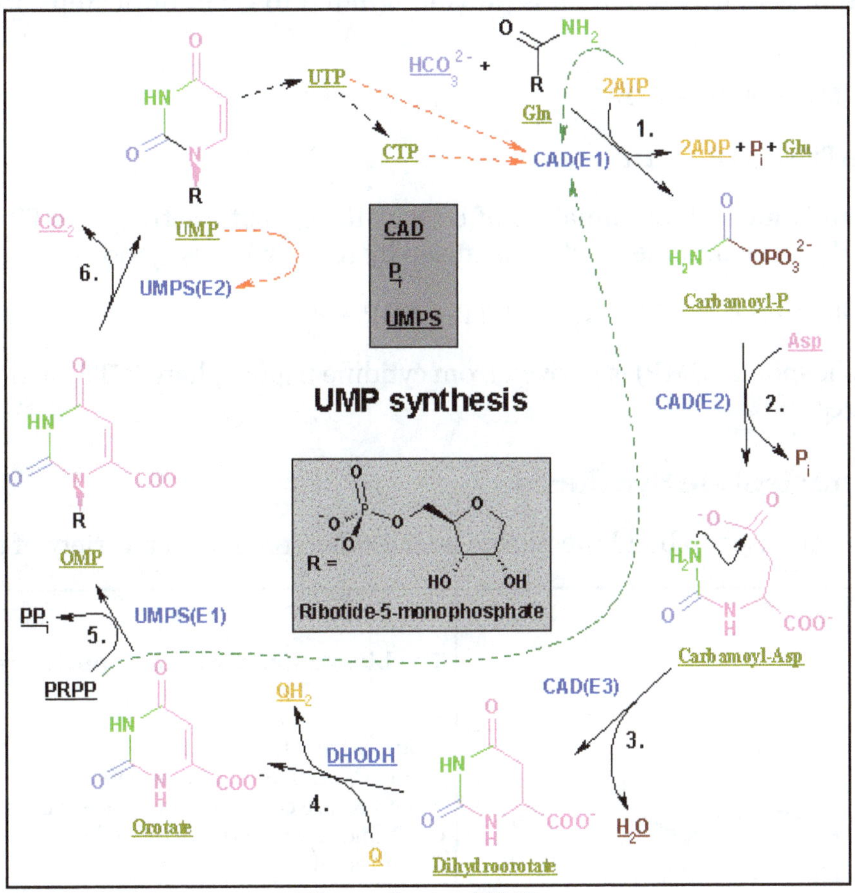

The synthesis of UMP.

The color scheme is as follows: enzymes, coenzymes, substrate names, inorganic molecules

The synthesis of the pyrimidines CTP and UTP occurs in the cytoplasm and starts with the formation of carbamoyl phosphate from glutamine and CO_2. Next, aspartate carbamoyltransferase catalyzes a condensation reaction between aspartate and carbamoyl phosphate to form carbamoyl aspartic acid, which is cyclized into 4,5-dihydroorotic acid by dihydroorotase. The latter is converted to orotate by dihydroorotate oxidase. The net reaction is:

(*S*)-Dihydroorotate + O_2 → Orotate + H_2O_2

Orotate is covalently linked with a phosphorylated ribosyl unit. The covalent linkage between the ribose and pyrimidine occurs at position C_1 of the ribose unit, which contains a pyrophosphate, and N_1 of the pyrimidine ring. Orotate phosphoribosyltransferase (PRPP transferase) catalyzes the net reaction yielding orotidine monophosphate (OMP):

Orotate + 5-Phospho-α-D-ribose 1-diphosphate (PRPP) → Orotidine 5'-phosphate + Pyrophosphate

Orotidine 5'-monophosphate is decarboxylated by orotidine-5'-phosphate decarboxylase to form uridine monophosphate (UMP). PRPP transferase catalyzes both the ribosylation and decarboxylation reactions, forming UMP from orotic acid in the presence of PRPP. It is from UMP that other pyrimidine nucleotides are derived. UMP is phosphorylated by two kinases to uridine triphosphate (UTP) via two sequential reactions with ATP. First the diphosphate form UDP is produced, which in turn is phosphorylated to UTP. Both steps are fueled by ATP hydrolysis:

ATP + UMP → ADP + UDP

UDP + ATP → UTP + ADP

CTP is subsequently formed by amination of UTP by the catalytic activity of CTP synthetase. Glutamine is the NH_3 donor and the reaction is fueled by ATP hydrolysis, too:

UTP + Glutamine + ATP + H_2O → CTP + ADP + P_i

Cytidine monophosphate (CMP) is derived from cytidine triphosphate (CTP) with subsequent loss of two phosphates.

Purine Ribonucleotide Synthesis

The atoms which are used to build the purine nucleotides come from a variety of sources:

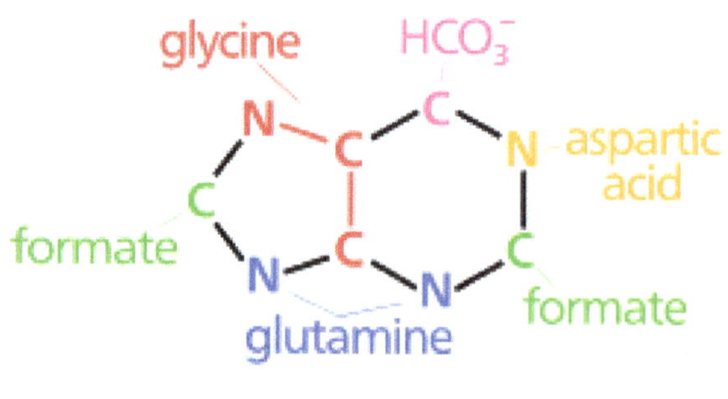

The biosynthetic origins of purine ring atoms

N_1 arises from the amine group of Asp
C_2 and C_8 originate from formate
N_3 and N_9 are contributed by the amide group of Gln
C_4, C_5 and N_7 are derived from Gly
C_6 comes from HCO_3^- (CO_2)

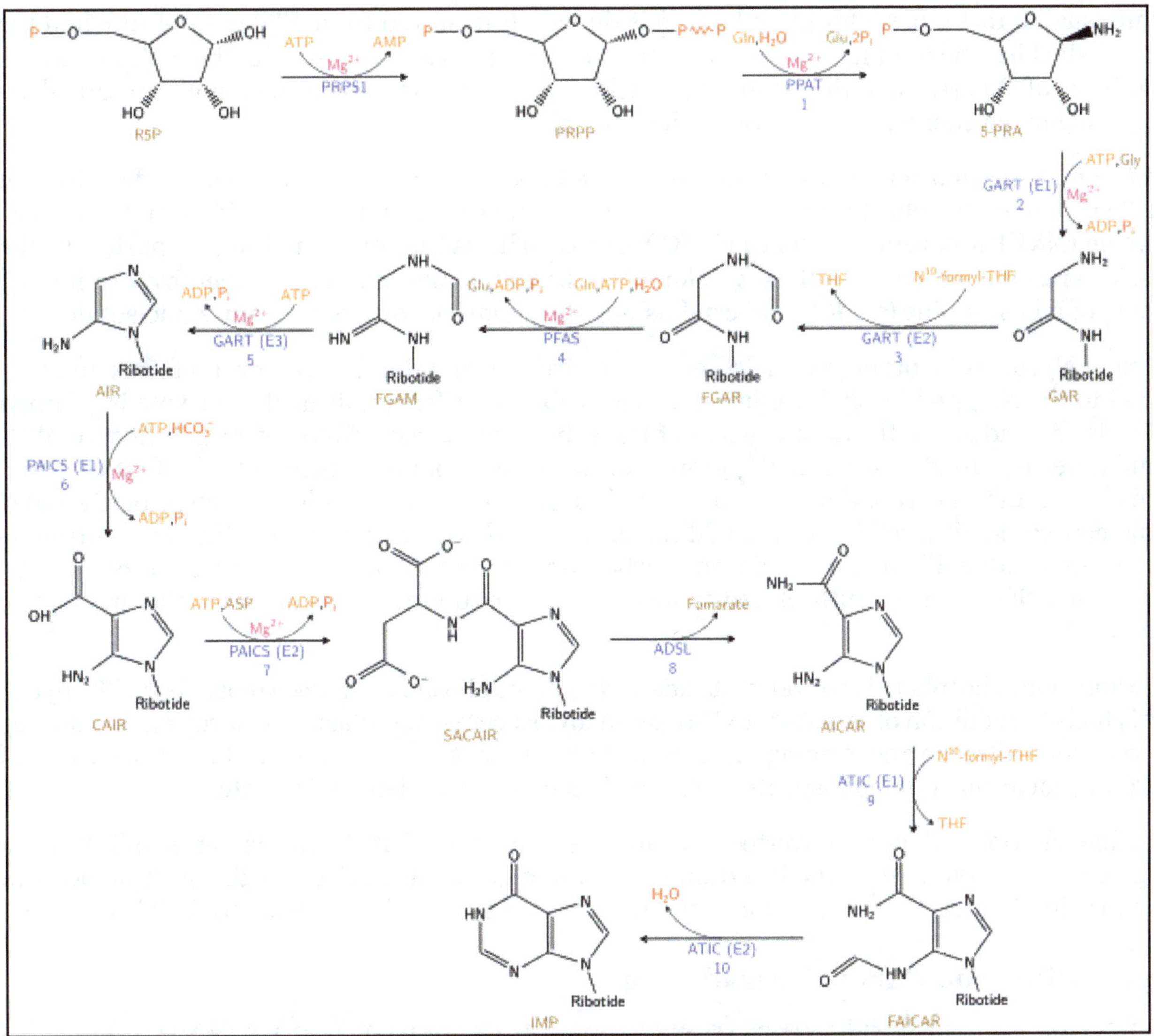

The synthesis of IMP. The color scheme is as follows: enzymes, coenzymes, substrate names, metal ions, inorganic molecules

The de novo synthesis of purine nucleotides by which these precursors are incorporated into the purine ring proceeds by a 10-step pathway to the branch-point intermediate IMP, the nucleotide of the base hypoxanthine. AMP and GMP are subsequently synthesized from this intermediate via separate, two-step pathways. Thus, purine moieties are initially formed as part of the ribonucleotides rather than as free bases.

Six enzymes take part in IMP synthesis. Three of them are multifunctional:

- GART (reactions 2, 3, and 5).
- PAICS (reactions 6, and 7).
- ATIC (reactions 9, and 10).

The pathway starts with the formation of PRPP. PRPS1 is the enzyme that activates R5P, which is formed primarily by the pentose phosphate pathway, to PRPP by reacting it with ATP. The reaction is unusual in that a pyrophosphoryl group is directly transferred from ATP to C_1 of R5P and that the product has the α configuration about C1. This reaction is also shared with the pathways for the synthesis of Trp, His, and the pyrimidine nucleotides. Being on a major metabolic crossroad and requiring much energy, this reaction is highly regulated.

In the first reaction unique to purine nucleotide biosynthesis, PPAT catalyzes the displacement of PRPP's pyrophosphate group (PP_i) by an amide nitrogen donated from either glutamine (N), glycine (N&C), aspartate (N), folic acid (C_1), or CO_2. This is the committed step in purine synthesis. The reaction occurs with the inversion of configuration about ribose C_1, thereby forming β-5-phosphorybosylamine (5-PRA) and establishing the anomeric form of the future nucleotide.

Next, a glycine is incorporated fueled by ATP hydrolysis and the carboxyl group forms an amine bond to the NH_2 previously introduced. A one-carbon unit from folic acid coenzyme N_{10}-formyl-THF is then added to the amino group of the substituted glycine followed by the closure of the imidazole ring. Next, a second NH_2 group is transferred from a glutamine to the first carbon of the glycine unit. A carboxylation of the second carbon of the glycin unit is concomittantly added. This new carbon is modified by the additional of a third NH_2 unit, this time transferred from an aspartate residue. Finally, a second one-carbon unit from formyl-THF is added to the nitrogen group and the ring covalently closed to form the common purine precursor inosine monophosphate (IMP).

Inosine monophosphate is converted to adenosine monophosphate in two steps. First, GTP hydrolysis fuels the addition of aspartate to IMP by adenylosuccinate synthase, substituting the carbonyl oxygen for a nitrogen and forming the intermediate adenylosuccinate. Fumarate is then cleaved off forming adenosine monophosphate. This step is catalyzed by adenylosuccinate lyase.

Inosine monophosphate is converted to guanosine monophosphate by the oxidation of IMP forming xanthylate, followed by the insertion of an amino group at C_2. NAD^+ is the electron acceptor in the oxidation reaction. The amide group transfer from glutamine is fueled by ATP hydrolysis.

Pyrimidine and Purine Degradation

In humans, pyrimidine rings (C, T, U) can be degraded completely to CO_2 and NH_3 (urea excretion). That having been said, purine rings (G, A) cannot. Instead they are degraded to the metabolically inert uric acid which is then excreted from the body. Uric acid is formed when GMP is split into the base guanine and ribose. Guanine is deaminated to xanthine which in turn is oxidized to uric acid. This last reaction is irreversible. Similarly, uric acid can be formed when AMP is deaminated to IMP from which the ribose unit is removed to form hypoxanthine. Hypoxanthine is oxidized to xanthine and finally to uric acid. Instead of uric acid secretion, guanine and IMP can be used for recycling purposes and nucleic acid synthesis in the presence of PRPP and aspartate (NH_3 donor).

Unnatural Base Pair (UBP)

An unnatural base pair (UBP) is a designed subunit (or nucleobase) of DNA which is created in a laboratory and does not occur in nature. In 2012, a group of American scientists led by Floyd Romesberg, a chemical biologist at the Scripps Research Institute in San Diego, California, published that his team designed an unnatural base pair (UBP). The two new artificial nucleotides or *Unnatural Base Pair* (UBP) were named d5SICS and dNaM. More technically, these artificial nucleotides bearing hydrophobic nucleobases, feature two fused aromatic rings that form a (d5SICS–dNaM) complex or base pair in DNA. In 2014 the same team from the Scripps Research Institute reported that they synthesized a stretch of circular DNA known as a plasmid containing natural T-A and C-G base pairs along with the best-performing UBP Romesberg's laboratory had designed, and inserted it into cells of the common bacterium *E. coli* that successfully replicated the unnatural base pairs through multiple generations. This is the first known example of a living organism passing along an expanded genetic code to subsequent generations. This was in part achieved by the addition of a supportive algal gene that expresses a nucleotide triphosphate transporter which efficiently imports the triphosphates of both d5SICSTP and dNaMTP into *E. coli* bacteria. Then, the natural bacterial replication pathways use them to accurately replicate the plasmid containing d5SICS–dNaM.

The successful incorporation of a third base pair is a significant breakthrough toward the goal of greatly expanding the number of amino acids which can be encoded by DNA, from the existing 20 amino acids to a theoretically possible 172, thereby expanding the potential for living organisms to produce novel proteins. The artificial strings of DNA do not encode for anything yet, but scientists speculate they could be designed to manufacture new proteins which could have industrial or pharmaceutical uses.

Length Unit

Nucleotide is a common unit of length for single-stranded nucleic acids, similar to how base pair is a unit of length for double-stranded nucleic acids.

Nucleotide Supplements

A study done by the Department of Sports Science at the University of Hull in Hull, UK has shown that nucleotides have significant impact on cortisol levels in saliva. Post exercise, the experimental nucleotide group had lower cortisol levels in their blood than the control or the placebo. Additionally, post supplement values of Immunoglobulin A were significantly higher than either the placebo or the control. The study concluded, "nucleotide supplementation blunts the response of the hormones associated with physiological stress."

Another study conducted in 2013 looked at the impact nucleotide supplementation had on the immune system in athletes. In the study, all athletes were male and were highly skilled in taekwondo. Out of the twenty athletes tested, half received a placebo and half received 480mg per day of nucleotide supplement. After thirty days, the study concluded that nucleotide supplementation may counteract the impairment of the body's immune function after heavy exercise.

Abbreviation Codes for Degenerate Bases

The IUPAC has designated the symbols for nucleotides. Apart from the five (A, G, C, T/U) bases, often degenerate bases are used especially for designing PCR primers. These nucleotide codes are listed here. Some primer sequences may also include the character "I", which codes for the non-standard nucleotide inosine. Inosine occurs in tRNAs, and will pair with adenine, cytosine, or thymine. This character does not appear in the following table however, because it does not represent a degeneracy. While inosine can serve a similar function as the degeneracy "D", it is an actual nucleotide, rather than a representation of a mix of nucleotides that covers each possible pairing needed.

<table>
<tr><th>Symbol</th><th>Description</th><th colspan="5">Bases represented</th></tr>
<tr><td>A</td><td>adenine</td><td>A</td><td></td><td></td><td></td><td rowspan="5">1</td></tr>
<tr><td>C</td><td>cytosine</td><td></td><td>C</td><td></td><td></td></tr>
<tr><td>G</td><td>guanine</td><td></td><td></td><td>G</td><td></td></tr>
<tr><td>T</td><td>thymine</td><td></td><td></td><td></td><td>T</td></tr>
<tr><td>U</td><td>uracil</td><td></td><td></td><td></td><td>U</td></tr>
<tr><td>W</td><td>weak</td><td>A</td><td></td><td></td><td>T</td><td rowspan="6">2</td></tr>
<tr><td>S</td><td>strong</td><td></td><td>C</td><td>G</td><td></td></tr>
<tr><td>M</td><td>amino</td><td>A</td><td>C</td><td></td><td></td></tr>
<tr><td>K</td><td>keto</td><td></td><td></td><td>G</td><td>T</td></tr>
<tr><td>R</td><td>purine</td><td>A</td><td></td><td>G</td><td></td></tr>
<tr><td>Y</td><td>pyrimidine</td><td></td><td>C</td><td></td><td>T</td></tr>
<tr><td>B</td><td>not A (B comes after A)</td><td></td><td>C</td><td>G</td><td>T</td><td rowspan="4">3</td></tr>
<tr><td>D</td><td>not C (D comes after C)</td><td>A</td><td></td><td>G</td><td>T</td></tr>
<tr><td>H</td><td>not G (H comes after G)</td><td>A</td><td>C</td><td></td><td>T</td></tr>
<tr><td>V</td><td>not T (V comes after T and U)</td><td>A</td><td>C</td><td>G</td><td></td></tr>
<tr><td>N</td><td>any base (not a gap)</td><td>A</td><td>C</td><td>G</td><td>T</td><td>4</td></tr>
</table>

Nucleic Acid

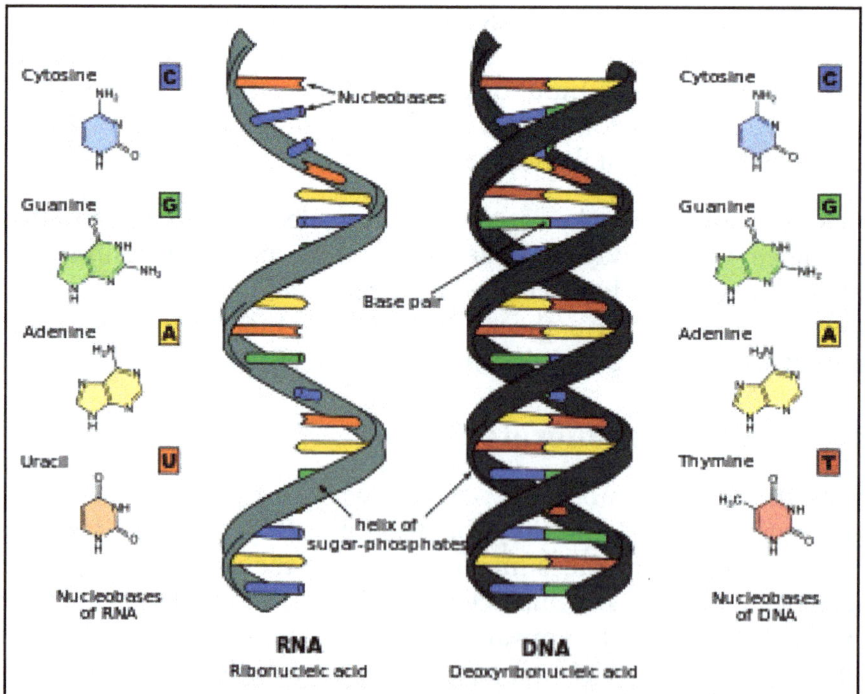

A comparison of the two principal nucleic acids: RNA (*left*) and DNA (*right*), showing the helices and nucleobases each employs.

The Swiss scientist Friedrich Miescher discovered nucleic acids (DNA) in 1869. Later, he raised the idea that they could be involved in heredity.

Nucleic acids are biopolymers, or large biomolecules, essential to all known forms of life. They are composed of monomers, which are nucleotides made of three components: a 5-carbon sugar, a phosphate group, and a nitrogenous base. If the sugar is a simple ribose, the polymer is RNA (ribonucleic acid); if the sugar is derived from ribose as deoxyribose, the polymer is DNA (deoxyribonucleic acid).

Nucleic acids are arguably the most important of all biomolecules. They are found in abundance in all living things, where they function to create and encode and then store information in the nucleus of every living cell of every life-form organism on Earth. In turn, they function to transmit and express that information inside and outside the cell nucleus—to the interior operations of the cell and ultimately to the next generation of each living organism. The encoded information is contained and conveyed via the nucleic acid sequence, which provides the 'ladder-step' ordering of nucleotides within the molecules of RNA and DNA.

Strings of nucleotides are bonded to form helical backbones—typically, one for RNA, two for DNA—and assembled into chains of base-pairs selected from the five primary, or canonical, nucleobases, which are: adenine, cytosine, guanine, thymine, and uracil; note, thymine occurs only in DNA and uracil only in RNA. Using amino acids and the process known as protein synthesis, the specific sequencing in DNA of these nucleobase-pairs enables storing and transmitting coded instructions as genes. In RNA, base-pair sequencing provides for manufacturing new proteins that determine the frames and parts and most chemical processes of all life forms.

History of Nucleic Acids

- Nucleic acids were discovered by Friedrich Miescher in 1869.

- In 1889 Richard Altmann discovered that nuclein has acidic properties, and it became called nucleic acid.
- In 1938 Astbury and Bell published the first X-ray diffraction pattern of DNA.
- In 1953 Watson and Crick determined the structure of DNA.

Experimental studies of nucleic acids constitute a major part of modern biological and medical research, and form a foundation for genome and forensic science, and the biotechnology and pharmaceutical industries.

Occurrence and Nomenclature

The term *nucleic acid* is the overall name for DNA and RNA, members of a family of biopolymers, and is synonymous with *polynucleotide*. Nucleic acids were named for their initial discovery within the nucleus, and for the presence of phosphate groups (related to phosphoric acid). Although first discovered within the nucleus of eukaryotic cells, nucleic acids are now known to be found in all life forms including within bacteria, archaea, mitochondria, chloroplasts, viruses, and viroids. All living cells contain both DNA and RNA (except some cells such as mature red blood cells), while viruses contain either DNA or RNA, but usually not both. The basic component of biological nucleic acids is the nucleotide, each of which contains a pentose sugar (ribose or deoxyribose), a phosphate group, and a nucleobase. Nucleic acids are also generated within the laboratory, through the use of enzymes (DNA and RNA polymerases) and by solid-phase chemical synthesis. The chemical methods also enable the generation of altered nucleic acids that are not found in nature, for example peptide nucleic acids.

Molecular Composition and Size

Nucleic acids are generally very large molecules. Indeed, DNA molecules are probably the largest individual molecules known. Well-studied biological nucleic acid molecules range in size from 21 nucleotides (small interfering RNA) to large chromosomes (human chromosome 1 is a single molecule that contains 247 million base pairs).

In most cases, naturally occurring DNA molecules are double-stranded and RNA molecules are single-stranded. There are numerous exceptions, however—some viruses have genomes made of double-stranded RNA and other viruses have single-stranded DNA genomes, and, in some circumstances, nucleic acid structures with three or four strands can form.

Nucleic acids are linear polymers (chains) of nucleotides. Each nucleotide consists of three components: a purine or pyrimidine nucleobase (sometimes termed *nitrogenous base* or simply *base*), a pentose sugar, and a phosphate group. The substructure consisting of a nucleobase plus sugar is termed a nucleoside. Nucleic acid types differ in the structure of the sugar in their nucleotides–DNA contains 2'-deoxyribose while RNA contains ribose (where the only difference is the presence of a hydroxyl group). Also, the nucleobases found in the two nucleic acid types are different: adenine, cytosine, and guanine are found in both RNA and DNA, while thymine occurs in DNA and uracil occurs in RNA.

The sugars and phosphates in nucleic acids are connected to each other in an alternating chain (sugar-phosphate backbone) through phosphodiester linkages. In conventional nomenclature, the carbons to which the phosphate groups attach are the 3'-end and the 5'-end carbons of the sugar. This gives nucleic acids directionality, and the ends of nucleic acid molecules are referred to as 5'-end and 3'-end. The nucleobases are joined to the sugars via an N-glycosidic linkage involving a nucleobase ring nitrogen (N-1 for pyrimidines and N-9 for purines) and the 1' carbon of the pentose sugar ring.

Non-standard nucleosides are also found in both RNA and DNA and usually arise from modification of the standard nucleosides within the DNA molecule or the primary (initial) RNA transcript. Transfer RNA (tRNA) molecules contain a particularly large number of modified nucleosides.

Topology

Double-stranded nucleic acids are made up of complementary sequences, in which extensive Watson-Crick base pairing results in a highly repeated and quite uniform double-helical three-dimensional structure. In contrast, single-stranded RNA and DNA molecules are not constrained to a regular double helix, and can adopt highly complex three-dimensional structures that are based on short stretches of intramolecular base-paired sequences including both Watson-Crick and noncanonical base pairs, and a wide range of complex tertiary interactions.

Nucleic acid molecules are usually unbranched, and may occur as linear and circular molecules. For example, bacterial chromosomes, plasmids, mitochondrial DNA, and chloroplast DNA are usually circular double-stranded DNA molecules, while chromosomes of the eukaryotic nucleus are usually linear double-stranded DNA molecules. Most RNA molecules are linear, single-stranded molecules, but both circular and branched molecules can result from RNA splicing reactions.

Nucleic Acid Sequences

One DNA or RNA molecule differs from another primarily in the sequence of nucleotides. Nucleotide sequences are of great importance in biology since they carry the ultimate instructions that encode all biological molecules, molecular assemblies, subcellular and cellular structures, organs, and organisms, and directly enable cognition, memory, and behavior. Enormous efforts have gone into the development of experimental methods to determine the nucleotide sequence of biological DNA and RNA molecules, and today hundreds of millions of nucleotides are sequenced daily at genome centers and smaller laboratories worldwide. In addition to maintaining the GenBank nucleic acid sequence database, the National Center for Biotechnology Information provides analysis and retrieval resources for the data in GenBank and other biological data made available through the NCBI Web site.

Types of Nucleic Acids

Deoxyribonucleic Acid

Deoxyribonucleic acid (DNA) is a nucleic acid containing the genetic instructions used in the development and functioning of all known living organisms. The DNA segments carrying this genetic information are called genes. Likewise, other DNA sequences have structural purposes, or are involved in regulating the use of this genetic information. Along with RNA and proteins, DNA is one of the three major macromolecules that are essential for all known forms of life. DNA consists of two long polymers of simple units called nucleotides, with backbones made of sugars and phosphate groups joined by ester bonds. These two strands run in opposite directions to each other and are, therefore, anti-parallel. Attached to each sugar is one of four types of molecules called nucleobases (informally, bases). It is the sequence of these four nucleobases along the backbone that encodes information. This information is read using the genetic code, which specifies the sequence of the amino acids within proteins. The code is read by copying stretches of DNA into the related nucleic acid RNA in a process called transcription. Within cells DNA is organized into long structures called chromosomes. During cell division these chromosomes are duplicated in the process of DNA replication, providing each cell its own complete set of chromosomes. Eukaryotic organisms (animals, plants, fungi, and protists) store most of their DNA inside the cell nucleus and some of their DNA in organelles, such as mitochondria or chloroplasts. In contrast, prokaryotes (bacteria and archaea) store their DNA only in the cytoplasm. Within the chromosomes, chromatin proteins such as histones compact and organize DNA. These compact structures guide the interactions between DNA and other proteins, helping control which parts of the DNA are transcribed.

Ribonucleic Acid

Ribonucleic acid (RNA) functions in converting genetic information from genes into the amino acid sequences of proteins. The three universal types of RNA include transfer RNA (tRNA), messenger RNA (mRNA), and ribosomal RNA (rRNA). Messenger RNA acts to carry genetic sequence information between DNA and ribosomes, directing protein synthesis. Ribosomal RNA is a major component of the ribosome, and catalyzes peptide bond formation. Transfer RNA serves as the carrier molecule for amino acids to be used in protein synthesis, and is responsible for decoding the mRNA. In addition, many other classes of RNA are now known.

Artificial Nucleic Acid

Artificial nucleic acid analogues have been designed and synthesized by chemists, and include peptide nucleic acid, morpholino- and locked nucleic acid, glycol nucleic acid, and threose nucleic acid. Each of these is distinguished from naturally occurring DNA or RNA by changes to the backbone of the molecules.

Carbohydrates

Carbohydrates are the polyhydroxy aldehydes or ketones. Aldehyde sugars are often referred to as aldoses while ketone sugars are called ketoses. Monosaccharides such as glucose and fructose are the simplest carbohydrates as given in the figure below. The monosaccharides can join covalently to give disaccharides, oligosaccharides, and polysaccharides. Unlike proteins and nucleic acid, polysachharides can be branched. Based on their functions, polysaccharides can be classified as structural polysaccharides (*e.g.* cellulose and chitin) and storage polysachharides (*e.g.* starch and glycogen).

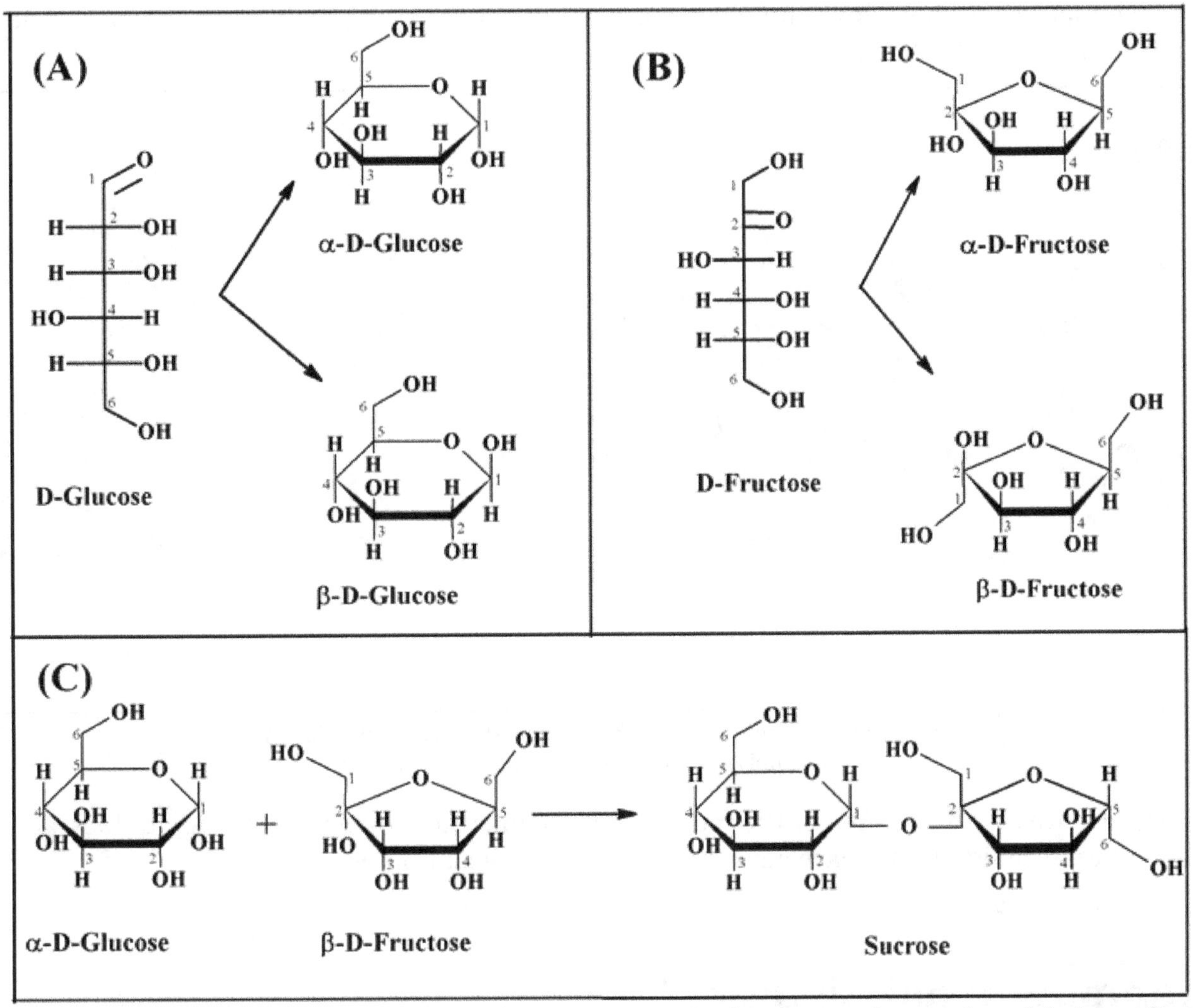

Open chain and ring structures of glucose (A) and fructose (B). Notice the two stereoisomers formed during cyclization of the open chain structure. A disaccharide (sucrose) formed from condensation of glucose and fructose (C).

Lipids

Lipids are amphipathic molecules with polar head groups and non-polar hydrocarbon region. Like carbohydrates, lipids also have both structural and storage roles in living systems. Apart from these, lipids play important roles in signal transduction pathways inside the cells. In vertebrates, for example, triacylglycerols are stored as fuels in specialized cells called adipocytes. Phosphoinositides, phosphorylated forms of phosphatidylinositol , are involved in cell signaling and membrane trafficking. Structural lipids perhaps constitute the most important class of the lipids because it is the structural lipids that define a cell. (Figure A)shows the structure of a glycerophospholipid, the most common structural lipid present in biomembranes.

The amphipathic nature of the lipids allows them to self-assemble in aqueous solutions to form a bilayer (Figure B). The assembled lipid bilayer has polar head groups interacting with water molecules whereas non-polar hydrocarbon regions are buried inside making the membrane core highly hydrophobic that acts as a barrier for polar and charged chemical species. It is this barrier that separates a cell from the external environment. A cell, however, needs to communicate and exchange material with its environment. To accomplish these tasks, biological membranes have acquired specialized proteins.

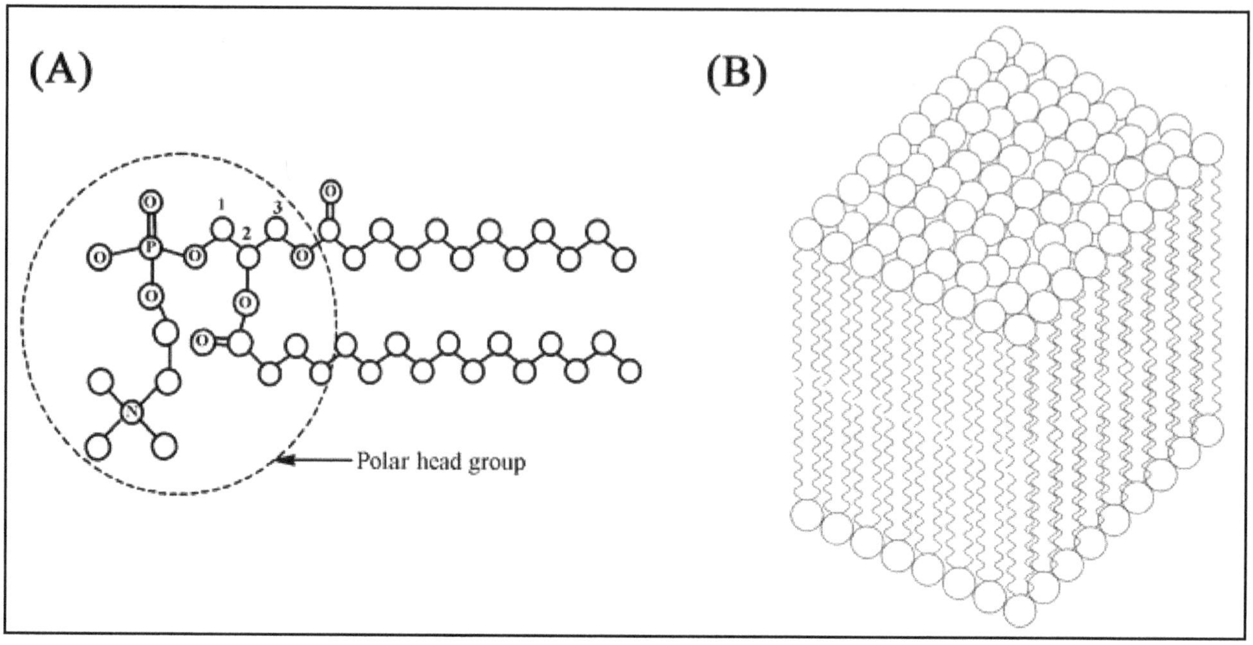

A glycerophospholipid showing the amphipathic nature of lipids (A).
Self-assembly of lipids producing a lipid bilayer (B).

Having reviewed the structures of the four major biomolecules, we are now ready to look at the bioanalytical techniques that have gained importance in recent times.

Computational Biology

Computational biology involves the development and application of data-analytical and theoretical methods, mathematical modeling and computational simulation techniques to the study of biological, behavioral, and social systems. The field is broadly defined and includes foundations in computer science, applied mathematics, animation, statistics, biochemistry, chemistry, biophysics, molecular biology, genetics, genomics, ecology, evolution, anatomy, neuroscience, and visualization.

Computational biology is different from biological computation, which is a subfield of computer science and computer engineering using bioengineering and biology to build computers, but is similar to bioinformatics, which is an interdisciplinary science using computers to store and process biological data.

Introduction

Computational Biology, sometimes referred to as bioinformatics, is the science of using biological data to develop algorithms and relations among various biological systems. Prior to the advent of computational biology, biologists did not have access to large amounts of data. Researchers were able to develop analytical methods for interpreting biological information, but were unable to share them quickly among colleagues.

Bioinformatics began to develop in the early 1970s. It was considered the science of analyzing informatics processes of various biological systems. At this time, research in artificial intelligence was using network models of the human brain in order to generate new algorithms. This use of biological data to develop other fields pushed biological researchers to revisit the idea of using computers to evaluate and compare large data sets. By 1982, information was being shared amongst researchers through the use of punch cards. The amount of data being shared began to grow exponentially by the end of the 1980s. This required the development of new computational methods in order to quickly analyze and interpret relevant information.

Since the late 1990s, computational biology has become an important part of developing emerging technologies for the field of biology. The terms computational biology and evolutionary computation have a similar name, but are not to be confused. Unlike computational biology, evolutionary computation is not concerned with modeling and analyzing biological data. It instead creates algorithms based on the ideas of evolution across species. Sometimes referred to as genetic algorithms, the research of this field can be applied to computational biology. While evolutionary computation is not inherently a part of computational biology, Computational evolutionary biology is a subfield of it.

Computational biology has been used to help sequence the human genome, create accurate models of the human brain, and assist in modeling biological systems.

Subfields

Computational Anatomy

Computational anatomy is a discipline focusing on the study of anatomical shape and form at the visible or gross anatomical scale of morphology. It involves the development and application of computational, mathematical and data-analytical methods for modeling and simulation of biological structures. The field is broadly defined and includes foundations in anatomy, applied mathematics and pure mathematics, machine learning, computational mechanics, computational science, medical imaging, neuroscience, physics, probability, and statistics; it also has strong connections with fluid mechanics and geometric mechanics. It focuses on the anatomical structures being imaged, rather than the medical imaging devices. It is similar in spirit to the history of Computational linguistics, a discipline that focuses on the linguistic structures rather than the sensor acting as the transmission and communication medium(s).Due to the availability of dense 3D measurements via technologies such as magnetic resonance imaging (MRI), Computational anatomy has emerged as a subfield of medical imaging and bioengineering for extracting anatomical coordinate systems at the morphome scale in 3D.

In computational anatomy, the diffeomorphism group is used to study different coordinate systems via coordinate transformations as generated via the Lagrangian and Eulerian velocities of flow from one anatomical configuration in to another. Computational anatomy intersects the study of Riemannian manifolds where groups of diffeomorphisms are the central focus, intersecting with emerging high-dimensional theories of shape emerging from the field of shape statistics.

The metric structures in Computational anatomy are related in spirit to morphometrics, with the distinction that Computational anatomy focuses on an infinite-dimensional space of coordinate systems transformed by a diffeomorphism, hence the central use of the terminology diffeomorphometry, the metric space study of coordinate systems via diffeomorphisms. At Computational anatomy's heart is the comparison of shape by recognizing in one shape the other. This connects it to D'Arcy Wentworth Thompson's developments On Growth and Form which has led to scientific explanations of morphogenesis, the process by which patterns are formed in Biology. The original formulation of Computational anatomy is as a generative model of shape and form from exemplars acted upon via transformations.

The spirit of this discipline shares strong overlap with areas such as computer vision and kinematics of rigid bodies, where objects are studied by analysing the groups responsible for the movement in question. It is a branch of the image analysis and pattern theory school at Brown University pioneered by Ulf Grenander. making spaces of anatomical patterns in Pattern Theory, into a metric space is one of the fundamental operations since being able to cluster and recognize anatomical configurations often requires a metric of close and far between shapes. The diffeomorphometry metric of Computational anatomy measures how far two diffeomorphic changes of coordinates are from each other, which in turn induces a metric on the shapes and images indexed to them. The models of metric pattern theory, in particular group action on the orbit of shapes and forms is a central tool to the formal definitions in Computational anatomy.

Computational Biomodeling

Computational biomodeling is a field concerned with building computer models of biological systems. Computational biomodeling aims to develop and use visual simulations in order to assess the complexity of biological systems. This is accomplished through the use of specialized algorithms, and visualization software. These models allow for prediction of how systems will react under different environments. This is useful for determining if a system is robust. A robust biological system is one that "maintain their state and functions against external and internal perturbations", which is essential for a biological system to survive. Computational biomodeling generates a large archive of such data, allowing for analysis from multiple users. While current techniques focus on small biological systems, researchers are working on approaches that will allow for larger networks to be analyzed and modeled. A majority of researchers believe that this will be essential in developing modern medical approaches to creating new drugs and gene therapy. A useful modelling approach is to use Petri nets via tools such as esyN.

Computational Genomics (Computational Genetics)

Computational genomics is a field within genomics which studies the genomes of cells and organisms. It is often referred to as Computational and Statistical Genetics. The Human Genome Project is one example of computational genomics. This project looks to sequence the entire human genome into a set of data. Once fully implemented, this could allow for doctors to analyze the genome of an individual patient. This opens the possibility of personalized medicine, prescribing treatments based on an individual's pre-existing genetic patterns. This project has created many similar programs. Researchers are looking to sequence the genomes of animals, plants, bacteria, and all other types of life.

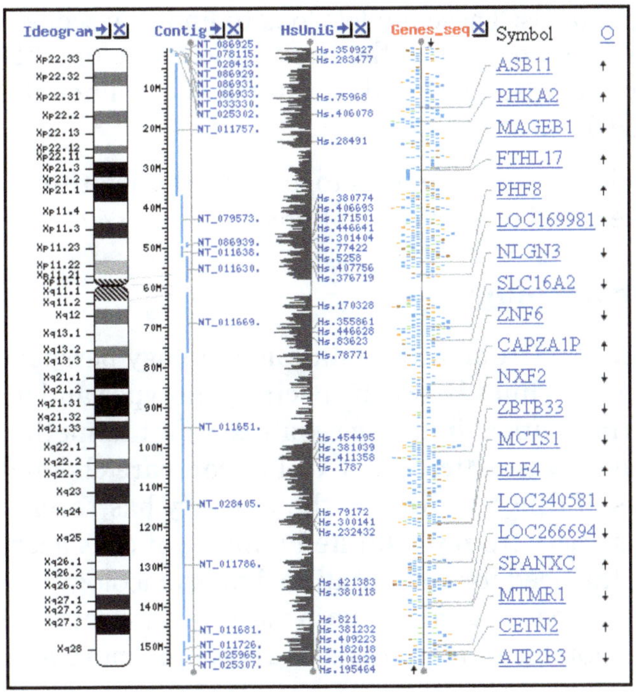

A partially sequenced genome.

One of the main ways that genomes are compared is by homology. Homology is the study of biological structures and nucleotide sequences in different organisms that come from a common ancestor. Research suggests that between 80 and 90% of genes in newly sequenced prokaryotic genomes can be identified this way.

This field is still in development. An untouched project in the development of computational genomics is the analysis of intergenic regions. Studies show that roughly 97% of the human genome consists of these regions. Researchers in computational genomics are working on understanding the functions of non-coding regions of the human genome through the development of computational and statistical methods and via large consortia projects such as ENCODE (The Encyclopedia of DNA Elements) and the Roadmap Epigenomics Project.

Computational Neuroscience

Computational neuroscience is the study of brain function in terms of the information processing properties of the structures that make up the nervous system. It is a subset of the field of neuroscience, and looks to analyze brain data to create practical applications. It looks to model the brain in order to examine specific types aspects of the neurological system. Various types of models of the brain include:

- Realistic Brain Models: These models look to represent every aspect of the brain, including as much detail at the cellular level as possible. Realistic models provide the most information about the brain, but also have the largest margin for error. More variables in a brain model create the possibility for more error to occur. These models do not account for parts of the cellular structure that scientists do not know about. Realistic brain models are the most computationally heavy and the most expensive to implement.

- Simplifying Brain Models: These models look to limit the scope of a model in order to assess a specific physical property of the neurological system. This allows for the intensive computational problems to be solved, and reduces the amount of potential error from a realistic brain model.

It is the work of computational neuroscientists to improve the algorithms and data structures currently used to increase the speed of such calculations.

Computational Pharmacology

Computational pharmacology (from a computational biology perspective) is "the study of the effects of genomic data to find links between specific genotypes and diseases and then screening drug data". The pharmaceutical industry requires a shift in methods to analyze drug data. Pharmacologists were able to use Microsoft Excel to compare chemical and genomic data related to the effectiveness of drugs. However, the industry has reached what is referred to as the Excel barricade. This arises from the limited number of cells accessible on a spreadsheet. This development led to the need for computational pharmacology. Scientists and researchers develop computational methods to analyze these massive data sets. This allows for an efficient comparison between the notable data points and allows for more accurate drugs to be developed.

Analysts project that if major medications fail due to patents, that computational biology will be necessary to replace current drugs on the market. Doctoral students in computational biology are being encouraged to pursue careers in industry rather than take Post-Doctoral positions. This is a direct result of major pharmaceutical companies needing more qualified analysts of the large data sets required for producing new drugs.

Computational Evolutionary Biology

Computational biology has assisted the field of evolutionary biology in many capacities. This includes:

- Using DNA data to reconstruct the tree of life with computational phylogenetics.
- Fitting population genetics models (either forward time or backward time) to DNA data to make inferences about demographic or selective history.
- Building population genetics models of evolutionary systems from first principles in order to predict what is likely to evolve.

Cancer Computational Biology

Cancer computational biology is a field that aims to determine the future mutations in cancer through an algorithmic approach to analyzing data. Research in this field has led to the use of high-throughput measurement. High throughput measurement allows for the gathering of millions of data points using robotics and other sensing devices. This data is collected from DNA, RNA, and other biological structures. Areas of focus include determining the characteristics of tumors, analyzing molecules that are deterministic in causing cancer, and understanding how the human genome relates to the causation of tumors and cancer.

Software and Tools

Computational Biologists use a wide range of software. These range from command line programs to graphical and web-based programs.

Open Source Software

Open source software provides a platform to develop computational biological methods. Specifically, open source means that anybody can access software developed in research. PLOS cites four main reasons for the use of open source software including:

- Reproducibility: This allows for researchers to use the exact methods used to calculate the relations between biological data.
- Faster Development: developers and researchers do not have to reinvent existing code for minor tasks. Instead they can use pre-existing programs to save time on the development and implementation of larger projects.
- Increased quality: Having input from multiple researchers studying the same topic provides a layer of assurance that errors will not be in the code.
- Long-term availability: Open source programs are not tied to any businesses or patents. This allows for them to be posted to multiple web pages and ensure that they are available in the future.

Computational Phylogenetics

Computational phylogenetics is the application of computational algorithms, methods, and programs to phylogenetic analyses. The goal is to assemble a phylogenetic tree representing a hypothesis about the evolutionary ancestry of a set of genes, species, or other taxa. For example, these techniques have been used to explore the family tree of hominid species and the relationships between specific genes shared by many types of organisms. Traditional phylogenetics relies on morphological data obtained by measuring and quantifying the phenotypic properties of representative organisms, while the more recent field of molecular phylogenetics uses nucleotide sequences encoding genes or amino acid sequences encoding proteins as the basis for classification. Many forms of molecular phylogenetics are closely related to and make extensive use of sequence alignment in constructing and refining phylogenetic trees, which are used to classify the evolutionary relationships between homologous genes represented in the genomes of divergent species. The phylogenetic trees constructed by computational methods are unlikely to perfectly reproduce the evolutionary tree that represents the historical relationships between the species being analyzed. The historical species tree may also differ from the historical tree of an individual homologous gene shared by those species.

Producing a phylogenetic tree requires a measure of homology among the characteristics shared by the taxa being compared. In morphological studies, this requires explicit decisions about which physical characteristics to measure and how to use them to encode distinct states corresponding to the input taxa. In molecular studies, a primary problem is in producing a multiple sequence alignment (MSA) between the genes or amino acid sequences of interest. Progressive sequence alignment methods produce a phylogenetic tree by necessity because they incorporate new sequences into the calculated alignment in order of genetic distance.

Types of Phylogenetic Trees and Networks

Phylogenetic trees generated by computational phylogenetics can be either *rooted* or *unrooted* depending on the input data and the algorithm used. A rooted tree is a directed graph that explicitly identifies a most recent common ancestor (MRCA), usually an imputed sequence that is not represented in the input. Genetic distance measures can be used to plot a tree with the input sequences as leaf nodes and their distances from the root proportional to their genetic distance from the hypothesized MRCA. Identification of a root usually requires the inclusion in the input data of at least one "outgroup" known to be only distantly related to the sequences of interest. By contrast, unrooted trees plot the distances and relationships between input sequences without making assumptions regarding their descent. An unrooted tree can always be produced from a rooted tree, but a root cannot usually be placed on an unrooted tree without additional data on divergence rates, such as the assumption of the molecular clock hypothesis.

The set of all possible phylogenetic trees for a given group of input sequences can be conceptualized as a discretely defined multidimensional "tree space" through which search paths can be traced by optimization algorithms. Although counting the total number of trees for a nontrivial number of input sequences can be complicated by variations in the definition of a tree topology, it is always true that there are more rooted than unrooted trees for a given number of inputs and choice of parameters.

Both rooted and unrooted phylogenetic trees can be further generalized to rooted or unrooted phylogenetic networks, which allow for the modeling of evolutionary phenomena such as hybridization or horizontal gene transfer.

Coding Characters and Defining Homology

Morphological Analysis

The basic problem in morphological phylogenetics is the assembly of a matrix representing a mapping from each of the taxa being compared to representative measurements for each of the phenotypic characteristics being used as a classifier. The types of phenotypic data used to construct this matrix depend on the taxa being compared; for individual species, they may involve measurements of average body size, lengths or sizes of particular bones or other physical features, or even behavioral manifestations. Of course, since not every possible phenotypic characteristic could be measured and encoded for analysis, the selection of which features to measure is a major inherent obstacle to the method. The decision of which traits to use as a basis for the matrix necessarily represents a hypothesis about which traits of a species or higher taxon are evolutionarily relevant. Morphological studies can be confounded by examples of convergent evolution of phenotypes. A major challenge in constructing useful classes is the high likelihood of inter-taxon overlap in the distribution of the phenotype's variation. The inclusion of extinct taxa in morphological analysis is often difficult due to absence of or incomplete fossil records, but has been shown to have a significant effect on the trees produced; in one study only the inclusion of extinct species of apes produced a morphologically derived tree that was consistent with that produced from molecular data.

Some phenotypic classifications, particularly those used when analyzing very diverse groups of taxa, are discrete and unambiguous; classifying organisms as possessing or lacking a tail, for example, is straightforward in the majority of cases, as is counting features such as eyes or vertebrae. However, the most appropriate representation of continuously varying phenotypic measurements is a controversial problem without a general solution. A common method is simply to sort the measurements of interest into two or more classes, rendering continuous observed variation as discretely classifiable (e.g., all examples with humerus bones longer than a given cutoff are scored as members of one state, and all members whose humerus bones are shorter than the cutoff are scored as members of a second state). This results in an easily manipulated data set but has been criticized for poor reporting of the basis for the class definitions and for sacrificing information compared to methods that use a continuous weighted distribution of measurements.

Because morphological data is extremely labor-intensive to collect, whether from literature sources or from field observations, reuse of previously compiled data matrices is not uncommon, although this may propagate flaws in the original matrix into multiple derivative analyses.

Molecular Analysis

The problem of character coding is very different in molecular analyses, as the characters in biological sequence data are immediate and discretely defined - distinct nucleotides in DNA or RNA sequences and distinct amino acids in protein sequences. However, defining homology can be challenging due to the inherent difficulties of multiple sequence alignment. For a given gapped MSA, several rooted phylogenetic trees can be constructed that vary in their interpretations of which changes are "mutations" versus ancestral characters, and which events are insertion mutations or deletion mutations. For example, given only a pairwise alignment with a gap region, it is impossible to determine whether one sequence bears an insertion mutation or the other carries a deletion. The problem is magnified in MSAs with unaligned and nonoverlapping gaps. In practice, sizable regions of a calculated alignment may be discounted in phylogenetic tree construction to avoid integrating noisy data into the tree calculation.

Distance-matrix Methods

Distance-matrix methods of phylogenetic analysis explicitly rely on a measure of "genetic distance" between the sequences being classified, and therefore they require an MSA as an input. Distance is often defined as the fraction of mismatches at aligned positions, with gaps either ignored or counted as mismatches. Distance methods attempt to construct an all-to-all matrix from the sequence query set describing the distance between each sequence pair. From this is constructed a phylogenetic tree that places closely related sequences under the same interior node and whose branch lengths closely reproduce the observed distances between sequences. Distance-matrix methods may produce either rooted or unrooted trees, depending on the algorithm used to calculate them. They are frequently used as the basis for progressive and iterative types of multiple sequence alignments. The main disadvantage of distance-matrix methods is their inability to efficiently use information about local high-variation regions that appear across multiple subtrees.

UPGMA and WPGMA

The UPGMA (*Unweighted Pair Group Method with Arithmetic mean*) and WPGMA (*Weighted Pair Group Method with Arithmetic mean*) methods produce rooted trees and require a constant-rate assumption - that is, it assumes an ultrametric tree in which the distances from the root to every branch tip are equal.

Neighbor-joining

Neighbor-joining methods apply general cluster analysis techniques to sequence analysis using genetic distance as a clustering metric. The simple neighbor-joining method produces unrooted trees, but it does not assume a constant rate of evolution (i.e., a molecular clock) across lineages.

Fitch-Margoliash Method

The Fitch-Margoliash method uses a weighted least squares method for clustering based on genetic distance. Closely related sequences are given more weight in the tree construction process to correct for the increased inaccuracy in measuring distances between distantly related sequences. The distances used as input to the algorithm must be normalized to prevent large artifacts in computing relationships between closely related and distantly related groups. The distances calculated by this method must be linear; the linearity criterion for distances requires that the expected values of the branch lengths for two individual branches must equal the expected value of the sum of the two branch distances - a property that applies to biological sequences only when they have been corrected for the possibility of back mutations at individual sites. This correction is done through the use of a substitution matrix such as that derived from the Jukes-Cantor model of DNA evolution. The distance correction is only necessary in practice when the evolution rates differ among branches. Another modification of the algorithm can be helpful, especially in case of concentrated distances (please report to concentration of measure phenomenon and curse of dimensionality): that modification, described in, has been shown to improve the efficiency of the algorithm and its robustness.

The least-squares criterion applied to these distances is more accurate but less efficient than the neighbor-joining methods. An additional improvement that corrects for correlations between distances that arise from many closely related sequences in the data set can also be applied at increased computational cost. Finding the optimal least-squares tree with any correction factor is NP-complete, so heuristic search methods like those used in maximum-parsimony analysis are applied to the search through tree space.

Using Outgroups

Independent information about the relationship between sequences or groups can be used to help reduce the tree search space and root unrooted trees. Standard usage of distance-matrix methods involves the inclusion of at least one outgroup sequence known to be only distantly related to the sequences of interest in the query set. This usage can be seen as a type of experimental control. If the outgroup has been appropriately chosen, it will have a much greater genetic distance and thus a longer branch length than any other sequence, and it will appear near the root of a rooted tree. Choosing an appropriate outgroup requires the selection of a sequence that is moderately related to the sequences of interest; too close a relationship defeats the purpose of the outgroup and too distant adds noise to the analysis. Care should also be taken to avoid situations in which the species from which the sequences were taken are distantly related, but the gene encoded by the sequences is highly conserved across lineages. Horizontal gene transfer, especially between otherwise divergent bacteria, can also confound outgroup usage.

Maximum Parsimony

Maximum parsimony (MP) is a method of identifying the potential phylogenetic tree that requires the smallest total number of evolutionary events to explain the observed sequence data. Some ways of scoring trees also include a "cost" associated with particular types of evolutionary events and attempt to locate the tree with the smallest total cost. This is a useful approach in cases where not every possible type of event is equally likely - for example, when particular nucleotides or amino acids are known to be more mutable than others.

The most naive way of identifying the most parsimonious tree is simple enumeration - considering each possible tree in succession and searching for the tree with the smallest score. However, this is only possible for a relatively small number of sequences or species because the problem of identifying the most parsimonious tree is known to be NP-hard; consequently a number of heuristic search methods for optimization have been developed to locate a highly parsimonious tree, if not the best in the set. Most such methods involve a steepest descent-style minimization mechanism operating on a tree rearrangement criterion.

Branch and Bound

The branch and bound algorithm is a general method used to increase the efficiency of searches for near-optimal solutions of NP-hard problems first applied to phylogenetics in the early 1980s. Branch and bound is particularly well suited to phylogenetic tree construction because it inherently requires dividing a problem into a tree structure as it subdivides the problem space into smaller regions. As its name implies, it requires as input both a branching rule (in the case of phylogenetics, the addition of the next species or sequence to the tree) and a bound (a rule that excludes certain regions of the search space from consideration, thereby assuming that the optimal solution cannot occupy that region). Identifying a good bound is the most challenging aspect of the algorithm's application to phylogenetics. A simple way of defining the bound is a maximum number of assumed evolutionary changes allowed per tree. A set of criteria known as Zharkikh's rules severely limit the search space by defining characteristics shared by all candidate "most parsimonious" trees. The two most basic rules require the elimination of all but one redundant sequence (for cases where multiple observations have produced identical data) and the elimination of character sites at which two or more states do not occur in at least two species. Under ideal conditions these rules and their associated algorithm would completely define a tree.

Sankoff-Morel-Cedergren Algorithm

The Sankoff-Morel-Cedergren algorithm was among the first published methods to simultaneously produce an MSA and a phylogenetic tree for nucleotide sequences. The method uses a maximum parsimony calculation in conjunction with a scoring function that penalizes gaps and mismatches, thereby favoring the tree that introduces a minimal number of such events (an alternative view holds that the trees to be favored are those that maximize the amount of sequence similarity that can be interpreted as homology, a point of view that may lead to different optimal trees). The imputed sequences at the interior nodes of the tree are scored and summed over all the nodes in each possible tree. The lowest-scoring tree sum provides both an optimal tree and an optimal MSA given the scoring function. Because the method is highly computationally intensive, an approximate method in which initial guesses for the interior alignments are refined one node at a time. Both the full and the approximate version are in practice calculated by dynamic programming.

MALIGN and POY

More recent phylogenetic tree/MSA methods use heuristics to isolate high-scoring, but not necessarily optimal, trees. The MALIGN method uses a maximum-parsimony technique to compute a multiple alignment by maximizing a cladogram score, and its companion POY uses an iterative method that couples the optimization of the phylogenetic tree with improvements in the corresponding MSA. However, the use of these methods in constructing evolutionary hypotheses has been criticized as biased due to the deliberate construction of trees reflecting minimal evolutionary events. This, in turn, has been countered by the view that such methods should be seen as heuristic approaches to find the trees that maximize the amount of sequence similarity that can be interpreted as homology.

Maximum Likelihood

The maximum likelihood method uses standard statistical techniques for inferring probability distributions to assign probabilities to particular possible phylogenetic trees. The method requires a substitution model to assess the probability of particular mutations; roughly, a tree that requires more mutations at interior nodes to explain the observed phylogeny will be assessed as having a lower probability. This is broadly similar to the maximum-parsimony method, but maximum likelihood allows additional statistical flexibility by permitting varying rates of evolution across both lineages and sites. In fact, the method requires that evolution at different sites and along different lineages must be statistically independent. Maximum likelihood is thus well suited to the analysis of distantly related sequences, but it is believed to be computationally intractable to compute due to its NP-hardness.

The "pruning" algorithm, a variant of dynamic programming, is often used to reduce the search space by efficiently calculating the likelihood of subtrees. The method calculates the likelihood for each site in a "linear" manner, starting at a node whose only descendants are leaves (that is, the tips of the tree) and working backwards toward the "bottom" node in nested sets. However, the trees produced by the method are only rooted if the substitution model is irreversible, which is not generally true of biological systems. The search for the maximum-likelihood tree also includes a branch length optimization component that is difficult to improve upon algorithmically; general global optimization tools such as the Newton-Raphson method are often used.

Bayesian Inference

Bayesian inference can be used to produce phylogenetic trees in a manner closely related to the maximum likelihood methods. Bayesian methods assume a prior probability distribution of the possible trees, which may simply be the probability of any one tree among all the possible trees that could be generated from the data, or may be a more sophisticated estimate derived from the assumption that divergence events such as speciation occur as stochastic processes. The choice of prior distribution is a point of contention among users of Bayesian-inference phylogenetics methods.

Implementations of Bayesian methods generally use Markov chain Monte Carlo sampling algorithms, although the choice of move set varies; selections used in Bayesian phylogenetics include circularly permuting leaf nodes of a proposed tree at each step and swapping descendant subtrees of a random internal node between two related trees. The use of Bayesian methods in phylogenetics has been controversial, largely due to incomplete specification of the choice of move set, acceptance criterion, and prior distribution in published work. Bayesian methods are generally held to be superior to parsimony-based methods; they can be more prone to long-branch attraction than maximum likelihood techniques, although they are better able to accommodate missing data.

Whereas likelihood methods find the tree that maximizes the probability of the data, a Bayesian approach recovers a tree that represents the most likely clades, by drawing on the posterior distribution. However, estimates of the posterior probability of clades (measuring their 'support') can be quite wide of the mark, especially in clades that aren't overwhelmingly likely. As such, other methods have been put forwards to estimate posterior probability.

Model Selection

Molecular phylogenetics methods rely on a defined substitution model that encodes a hypothesis about the relative rates of mutation at various sites along the gene or amino acid sequences being studied. At their simplest, substitution models aim to correct for differences in the rates of transitions and transversions in nucleotide sequences. The use of substitution models is necessitated by the fact that the genetic distance between two sequences increases linearly only for a short time after the two sequences diverge from each other (alternatively, the distance is linear only shortly before coalescence). The longer the amount of time after divergence, the more likely it becomes that two mutations occur at the same nucleotide site. Simple genetic distance calculations will thus undercount the number of mutation events that have occurred in evolutionary history. The extent of this undercount increases with increasing time since divergence, which can lead to the phenomenon of long branch attraction, or the misassignment of two distantly related but convergently evolving sequences as closely related. The maximum parsimony method is particularly susceptible to this problem due to its explicit search for a tree representing a minimum number of distinct evolutionary events.

Types of Models

All substitution models assign a set of weights to each possible change of state represented in the sequence. The most common model types are implicitly reversible because they assign the same weight to, for example, a G>C nucleotide mutation as to a C>G mutation. The simplest possible model, the Jukes-Cantor model, assigns an equal probability to every possible change of state for a given nucleotide base. The rate of change between any two distinct nucleotides will be one-third of the overall substitution rate. More advanced models distinguish between transitions and transversions. The most general possible time-reversible model, called the GTR model, has six mutation rate parameters. An even more generalized model known as the general

12-parameter model breaks time-reversibility, at the cost of much additional complexity in calculating genetic distances that are consistent among multiple lineages. One possible variation on this theme adjusts the rates so that overall GC content - an important measure of DNA double helix stability - varies over time.

Models may also allow for the variation of rates with positions in the input sequence. The most obvious example of such variation follows from the arrangement of nucleotides in protein-coding genes into three-base codons. If the location of the open reading frame (ORF) is known, rates of mutation can be adjusted for position of a given site within a codon, since it is known that wobble base pairing can allow for higher mutation rates in the third nucleotide of a given codon without affecting the codon's meaning in the genetic code. A less hypothesis-driven example that does not rely on ORF identification simply assigns to each site a rate randomly drawn from a predetermined distribution, often the gamma distribution or log-normal distribution. Finally, a more conservative estimate of rate variations known as the covarion method allows autocorrelated variations in rates, so that the mutation rate of a given site is correlated across sites and lineages.

Choosing the Best Model

The selection of an appropriate model is critical for the production of good phylogenetic analyses, both because underparameterized or overly restrictive models may produce aberrant behavior when their underlying assumptions are violated, and because overly complex or overparameterized models are computationally expensive and the parameters may be overfit. The most common method of model selection is the likelihood ratio test (LRT), which produces a likelihood estimate that can be interpreted as a measure of "goodness of fit" between the model and the input data. However, care must be taken in using these results, since a more complex model with more parameters will always have a higher likelihood than a simplified version of the same model, which can lead to the naive selection of models that are overly complex. For this reason model selection computer programs will choose the simplest model that is not significantly worse than more complex substitution models. A significant disadvantage of the LRT is the necessity of making a series of pairwise comparisons between models; it has been shown that the order in which the models are compared has a major effect on the one that is eventually selected.

An alternative model selection method is the Akaike information criterion (AIC), formally an estimate of the Kullback–Leibler divergence between the true model and the model being tested. It can be interpreted as a likelihood estimate with a correction factor to penalize overparameterized models. The AIC is calculated on an individual model rather than a pair, so it is independent of the order in which models are assessed. A related alternative, the Bayesian information criterion (BIC), has a similar basic interpretation but penalizes complex models more heavily.

A comprehensive step-by-step protocol on constructing phylogenetic tree, including DNA/Amino Acid contiguous sequence assembly, multiple sequence alignment, model-test (testing best-fitting substitution models) and phylogeny reconstruction using Maximum Likelihood and Bayesian Inference, is available at Nature Protocol.

An non traditional way of evaluating the Pylogenetic Tree is to compare it with clustering result. One can use a Multidimensional Scaling technique, so called Interpolative Joining to do dimensionality reduction to visualize the clustering result for the sequences in 3D, and then map the phylogenetic tree onto the clustering result. A better tree usually has a higher correlation with the clustering result.

Evaluating Tree Support

As with all statistical analysis, the estimation of phylogenies from character data requires an evaluation of confidence. A number of methods exist to test the amount of support for a phylogenetic tree, either by evaluating the support for each sub-tree in the phylogeny (nodal support) or evaluating whether the phylogeny is significantly different from other possible trees (alternative tree hypothesis tests).

Nodal Support

The most common method for assessing tree support is to evaluate the statistical support for each node on the tree. Typically, a node with very low support is not considered valid in further analysis, and visually may be collapsed into a polytomy to indicate that relationships within a clade are unresolved.

Consensus Tree

Many methods for assessing nodal support involve consideration of multiple phylogenies. The consensus tree summarizes the nodes that are shared among a set of trees. In a strict consensus, only nodes found in every tree are shown, and the rest are collapsed into an unresolved polytomy. Less conservative methods, such as the majority-rule consensus tree, consider nodes that are supported by a given percentage of trees under consideration (such as at least 50%).

For example, in maximum parsimony analysis, there may be many trees with the same parsimony score. A strict consensus tree would show which nodes are found in all equally parsimonious trees, and which nodes differ. Consensus trees are also used to evaluate support on phylogenies reconstructed with Bayesian inference.

Bootstrapping and Jackknifing

In statistics, the bootstrap is a method for inferring the variability of data that has an unknown distribution using pseudoreplications of the original data. For example, given a set of 100 data points, a pseudoreplicate is a data set of the same size (100 points) randomly sampled from the original data, with replacement. That is, each original data point may be represented more than once in the pseudoreplicate, or not at all. Statistical support involves evaluation of whether the original data has similar properties to a large set of pseudoreplicates.

In phylogenetics, bootstrapping is conducted using the columns of the character matrix. Each pseudoreplicate contains the same number of species (rows) and characters (columns) randomly sampled from the original matrix, with replacement. A phylogeny is reconstructed from each pseudoreplicate, with the same methods used to reconstruct the phylogeny from the original data. For each node on the phylogeny, the nodal support is the percentage of pseudoreplicates containing that node.

The statistical rigor of the bootstrap test has been empirically evaluated using viral populations with known evolutionary histories, finding that 70% bootstrap support corresponds to a 95% probability that the clade exists. However, this was tested under ideal conditions (e.g. no change in evolutionary rates, symmetric phylogenies). In practice, values above 70% are generally supported and left to the researcher or reader to evaluate confidence. Nodes with support lower than 70% are typically considered unresolved.

Jackknifing in phylogenetics is a similar procedure, except the columns of the matrix are sampled without replacement. Pseudoreplicates are generated by randomly subsampling the data—for example, a "10% jackknife" would involve randomly sampling 10% of the matrix many times to evaluate nodal support.

Posterior Probability

Reconstruction of phylogenies using Bayesian inference generates a posterior distribution of highly probable trees given the data and evolutionary model, rather than a single "best" tree. The trees in the posterior distribution generally have many different topologies. Most Bayesian inference methods utilize a Markov-chain Monte Carlo iteration, and the initial steps of this chain are not considered reliable reconstructions of the phylogeny. Trees generated early in the chain are usually discarded as burn-in. The most common method of evaluating nodal support in a Bayesian phylogenetic analysis is to calculate the percentage of trees in the posterior distribution (post-burn-in) which contain the node.

The statistical support for a node in Bayesian inference is expected to reflect the probability that a clade really exists given the data and evolutionary model. Therefore, the threshold for accepting a node as supported is generally higher than for bootstrapping.

Step Counting Methods

Bremer support counts the number of extra steps needed to contradict a clade.

Shortcomings

These measures each have their weaknesses. For example, smaller or larger clades tend to attract larger support values than mid-sized clades, simply as a result of the number of taxa in them.

Bootstrap support can provide high estimates of node support as a result of noise in the data rather than the true existence of a clade.

Limitations and Workarounds

Ultimately, there is no way to measure whether a particular phylogenetic hypothesis is accurate or not, unless the true relationships among the taxa being examined are already known (which may happen with bacteria or viruses under laboratory conditions). The best result an empirical phylogeneticist can hope to attain is a tree with branches that are well supported by the available evidence. Several potential pitfalls have been identified:

Homoplasy

Certain characters are more likely to evolve convergently than others; logically, such characters should be given less weight in the reconstruction of a tree. Weights in the form of a model of evolution can be inferred from sets of molecular data, so that maximum likelihood or Bayesian methods can be used to analyze them. For molecular sequences, this problem is exacerbated when the taxa under study have diverged substantially. As time since the divergence of two taxa increase, so does the probability of multiple substitutions on the same site, or back mutations, all of which result in homoplasies. For morphological data, unfortunately, the only objective way to determine convergence is by the construction of a tree – a somewhat circular method. Even so, weighting homoplasious characters does indeed lead to better-supported trees. Further refinement can be brought by weighting changes in one direction higher than changes in another; for instance, the presence of thoracic wings almost guarantees placement among the pterygote insects because, although wings are often lost secondarily, there is no evidence that they have been gained more than once.

Horizontal Gene Transfer

In general, organisms can inherit genes in two ways: vertical gene transfer and horizontal gene transfer. Vertical gene transfer is the passage of genes from parent to offspring, and horizontal (also called lateral) gene transfer occurs when genes jump between unrelated organisms, a common phenomenon especially in prokaryotes; a good example of this is the acquired antibiotic resistance as a result of gene exchange between various bacteria leading to multi-drug-resistant bacterial species. There have also been well-documented cases of horizontal gene transfer between eukaryotes.

Horizontal gene transfer has complicated the determination of phylogenies of organisms, and inconsistencies in phylogeny have been reported among specific groups of organisms depending on the genes used to construct evolutionary trees. The only way to determine which genes have been acquired vertically and which horizontally is to parsimoniously assume that the largest set of genes that have been inherited together have been inherited vertically; this requires analyzing a large number of genes.

Hybrids, Speciation, Introgressions and Incomplete Lineage Sorting

The basic assumption underlying the mathematical model of cladistics is a situation where species split neatly in bifurcating fashion. While such an assumption may hold on a larger scale (bar horizontal gene transfer), speciation is often much less orderly. Research since the cladistic method was introduced has shown that hybrid speciation, once thought rare, is in fact quite common, particularly in plants. Also paraphyletic speciation is common, making the assumption of a bifurcating pattern unsuitable, leading to phylogenetic networks rather than trees. Introgression can also move genes between otherwise distinct species and sometimes even genera, complicating phylogenetic analysis based on genes. This phenomenon can contribute to "incomplete lineage sorting" and is thought to be a common phenomenon across a number of groups. In species level analysis this can be dealt with by larger sampling or better whole genome analysis. Often the problem is avoided by restricting the analysis to fewer, not closely related specimen.

Taxon Sampling

Owing to the development of advanced sequencing techniques in molecular biology, it has become feasible to gather large amounts of data (DNA or amino acid sequences) to infer phylogenetic hypotheses. For example, it is not rare to find studies with character matrices based on whole mitochondrial genomes (~16,000 nucleotides, in many animals). However, simulations have shown that it is more important to increase the number of taxa in the matrix than to increase the number of characters, because the more taxa there are, the more accurate and more robust is the resulting phylogenetic tree. This may be partly due to the breaking up of long branches.

Phylogenetic Signal

Another important factor that affects the accuracy of tree reconstruction is whether the data analyzed actually contain a useful phylogenetic signal, a term that is used generally to denote whether a character evolves slowly enough to have the same state in closely related taxa as opposed to varying randomly. Tests for phylogenetic signal exist.

Continuous Characters

Morphological characters that sample a continuum may contain phylogenetic signal, but are hard to code as discrete characters. Several methods have been used, one of which is gap coding, and there are variations on gap coding. In the original form of gap coding:

> group means for a character are first ordered by size. The pooled within-group standard deviation is calculated ... and differences between adjacent means ... are compared relative to this standard deviation. Any pair of adjacent means is considered different and given different integer scores ... if the means are separated by a "gap" greater than the within-group standard deviation ... times some arbitrary constant.

If more taxa are added to the analysis, the gaps between taxa may become so small that all information is lost. Generalized gap coding works around that problem by comparing individual pairs of taxa rather than considering one set that contains all of the taxa.

Missing Data

In general, the more data that are available when constructing a tree, the more accurate and reliable the resulting tree will be. Missing data are no more detrimental than simply having fewer data, although the impact is greatest when most of the missing data are in a small number of taxa. Concentrating the missing data across a small number of characters produces a more robust tree.

The Role of Fossils

Because many characters involve embryological, or soft-tissue or molecular characters that (at best) hardly ever fossilize, and the interpretation of fossils is more ambiguous than that of living taxa, extinct taxa almost invariably have higher proportions of missing data than living ones. However, despite these limitations, the inclusion of fossils is invaluable, as they can provide information in sparse areas of trees, breaking up long branches and constraining intermediate character states; thus, fossil taxa contribute as much to tree resolution as modern taxa. Fossils can also constrain the age of lineages and thus demonstrate how consistent a tree is with the stratigraphic record; stratocladistics incorporates age information into data matrices for phylogenetic analyses.

Approaches in Bioanalysis and Bioassays

Initial bioanalytical methods were not highly specific and were relatively insensitive as compared to the modern methods. The assays included colorimetric estimation of the compounds or simple bioassays, such as antibiotic estimation by quantifying their ability to inhibit microbial growth. Development of pharmacokinetics during 1930s started demanding for more specific and sensitive methods to accurately determine the concentrations of drugs and metabolites in biological specimens. Around this time, spectroscopic techniques such as UV/Visible spectroscopy, infrared spectroscopy, and chiroptical spectroscopy were seeing advancement but were largely restricted to the analysis of chemical compounds. Lack of sensitive instrumentation around that time further restricted their applications to biological samples that usually have low concentrations of molecules. Second half of the 20th century saw a rapid development in the instrumentation and development of new methodologies that eventually would find applications in life sciences and medicine. Liquid chromatography turned out to be a major advancement towards achieving sensitivity and power of resolving the closely-related metabolites. Reversed-phase chromatography, for example, has proved to be an excellent tool for resolving and analyzing the small molecules with excellent sensitivity. Electrophoresis is another powerful tool for analyzing and separating biomolecules. It has turned out to be an indispensable tool for analyzing nucleic acids. Integrity of isolated nucleic acids, cleavage of DNA molecules by restriction enzymes, mapping of restriction sites in a DNA molecule, and joining of two or more DNA fragments by ligases are some of the diverse applications of electrophoresis in a molecular genetics laboratory (figure below). DNA molecules differing in even one base pair can be separated by electrophoresis; this allows sequencing of DNA by Sanger›s method. Electrophoresis is also used to analyze proteins. Electrophoresis allows separation of proteins based on their isoelectric points. SDS-PAGE (Sodium dodecyl – polyacrylamide gel electrophoresis) of proteins separates the proteins based on their size and therefore allows determination of their molecular weights.

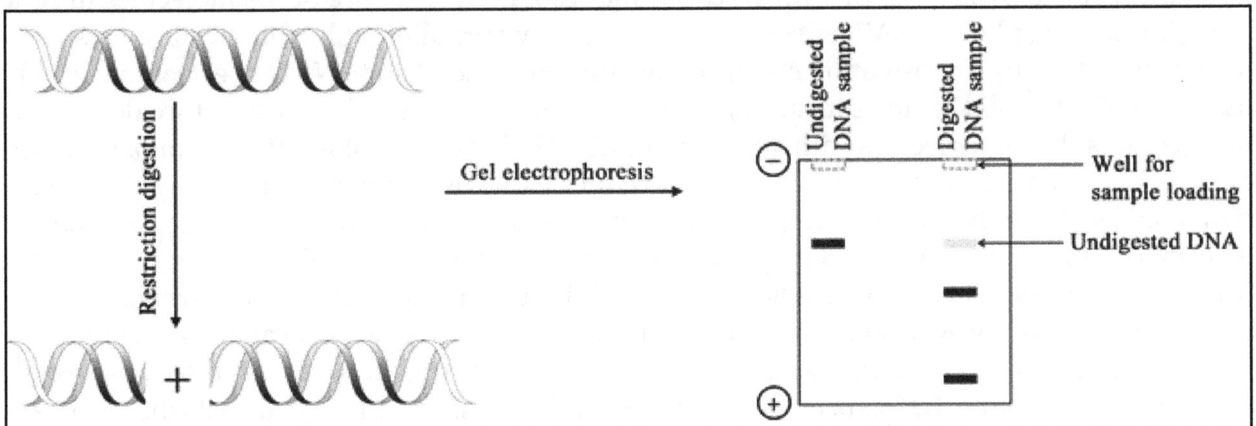

Restriction digestion of a DNA molecule assessed by agarose gel electrophoresis

Quantification of an analyte, is among the most common applications of analytical tools. You may be familiar with the use of UV/visible light for recording absorption of organic molecules to determine the concentration of the compound. It is therefore clear that light or electromagnetic radiation can interact with the matter providing useful information about it. Interaction of electromagnetic radiation with matter is termed as spectroscopy. Absorption of UV/visible radiation is associated with electronic transitions in the molecules; UV/Visible spectroscopy is therefore also referred to as the electronic spectroscopy. Absorption of ultraviolet and/or visible radiation is the most commonly employed method to estimate the concentration of biomolecules such as proteins, peptides, nucleotides, nucleic acids, carbohydrates, and lipids. Absorption at 260 nm and 280 nm provides information about the nucleic acid contamination in protein preparations. Phenol is commonly used to isolate nucleic acids; $\frac{A_{260}}{A_{280}}$ is used to determine phenol contamination in nucleic acid preparations and has become a routinely used method in molecular biology laboratories. Electronic spectroscopy goes beyond quantification of biomolecules: fluorescence spectroscopy is used to study various biological processes *viz.* protein folding/unfolding, binding studies, etc. Electronic circular dichroism spectroscopy is a chiroptical method and finds applications in analyzing protein and peptide structures, protein folding/unfolding, binding studies, etc.

Infrared spectroscopy probes the vibrational frequencies in the molecules; the frequency of vibration depends on the strength of the bond and the atoms involved thereby allowing identification of functional groups present in the organic molecules. As the absorption depends on the concentration, infrared spectroscopy can also be utilized for determining the concentrations of the analytes. The vibrational frequencies of the bonds are sensitive to the conformation of the molecule as well as the interactions of the atoms involved. Infrared spectroscopy can therefore provide information about the conformations of the molecules. In fact, infrared spectroscopy is often used to determine the secondary structures of the polypeptides. Advent of nuclear magnetic resonance (NMR) spectroscopy in 1940s revolutionized the analysis of small molecules. When used alongside infrared spectroscopy, NMR spectroscopy can quickly provide the complete structure of the molecules. Advancement in the hardware and development of the experimental methods has made NMR spectroscopy one of the most powerful weapons in a chemist's and biochemist's arsenal. NMR is routinely employed to study the structure and dynamics of biomacromolecules. In fact, NMR is the only tool that provides atomic resolution structure of the molecules in solution. This is a big plus for NMR spectroscopy over X-ray crystallography that needs a crystal for determining the atomic resolution structure. Furthermore, solid state NMR spectroscopy can be used to study the solid samples including single crystals. Atomic resolution structure determination requires the biomolecules with very high purity. High purity biological macromolecules are obtained through one or more chromatographic methods. The principle underlying the separation of molecules is their partitioning between a stationary and a mobile phase. The partition coefficient of a molecule depends on its physicochemical properties and molecules can be separated based on their size, charge, hydrophobicity or affinity to a particular ligand. Chromatographic techniques can also provide analytical information, *e.g.* molecular weight can be determined using size exclusion chromatography wherein there is a relationship between the molecular weight and the elution volume. NMR spectroscopy, however, has come a long way since its discovery and it is now possible to determine the structures of biomolecules in their native milieu *i.e.* inside the living cells.

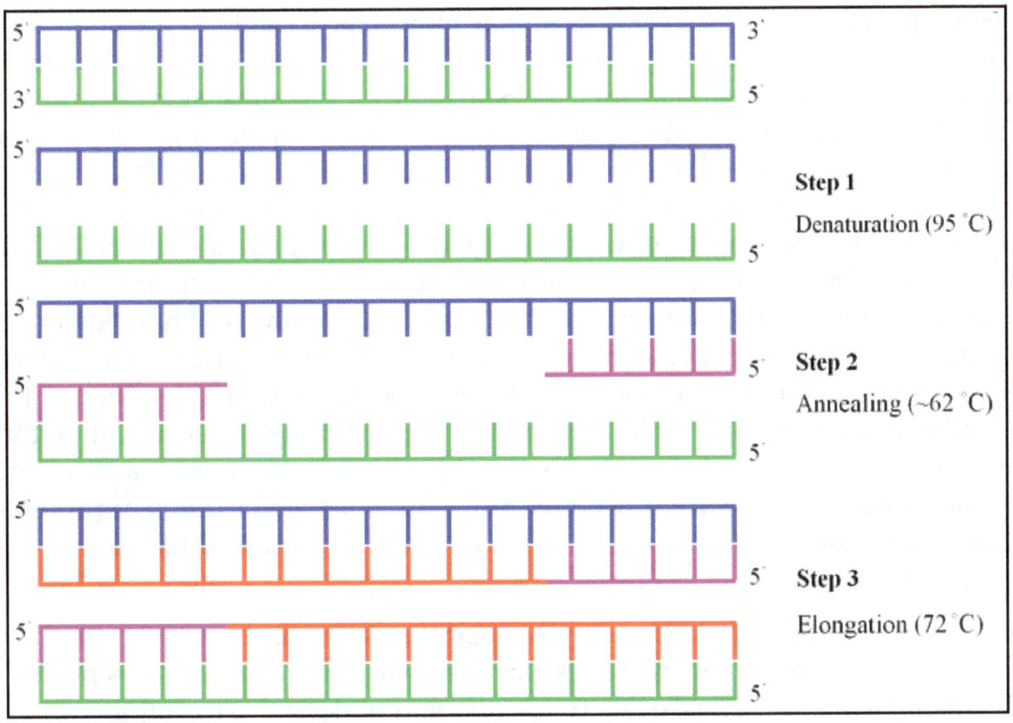

Principle of PCR amplification of DNA

Discovery of polymerase chain reaction (PCR) was a major step forward in the biomedical research and diagnostics. Presence of a pathogen inside the body is classically detected using serological methods or culture of the infectious agents. Owing to its excellent sensitivity, PCR can detect the presence of pathogens earlier than the serological tests. Other than infectious diseases, PCR is also used to detect genetic disorders. It is hard to imagine doing research in the areas of molecular genetics without employing PCR.

Complexity of the biological systems hardly needs any mention. To understand the molecules at function in a living system, it is important to look at the system altogether rather than individual components and processes. Sequencing of complete genomes led researchers to estimate the number of genes a particular organism expresses and further to understand the co-expression of a large number of genes and their role in physiology and pathophysiology. Identification and estimation of the subset of proteins expressed at any instant can provide useful information about the system *viz.* expression of a gene or a set of genes beyond a threshold level may be a marker of a disease. Intrinsically low levels of a large number of proteins, however, posed a challenge for detecting and identifying them. Application of mass spectrometry to proteins and peptides provided major breakthrough towards achieving this. Mild ionization techniques such as electrospray ionization (ESI) and matrix assisted laser desorption ionization (MALDI) could successfully ionize large biomolecules without damaging them. This opened up a plethora of possibilities and resulted in the development of a new research discipline, called *proteomics* . Proteomics refers to the study of the complete set of proteins expressed by a cell or an organism. The ultimate goal of the proteomic studies is to identify all the proteins present in the specimen; quantify them; and identify the posttranslational modifications, if any. A proteomic approach typically starts with the isolation of the total protein from the sample. Total protein is then resolved into its components using 2-D gel electrophoresis that separates the proteins according to their isoelectric points in one dimension and their molecular weights in the other. The individual protein spots are then cut from the gel and eluted out. The proteins are then analyzed using mass spectrometry either directly or after digesting with a sequence specific protease such as trypsin. The proteins can then be identified either by *de novo* sequencing or by using databases having sequence information and thereby mass information of the peptide fragments. Proteomic analysis is useful in identifying the markers for various processes and diseases. For example, comparing the samples from a set of healthy individuals with that of individuals having some disease/disorder can identify if the protein levels go up or down in the unhealthy individuals as compared to the healthy ones. Systematic studies with a large number of individuals are likely to result in identification of biomarkers for the diseases. The need to retrieve and analyze the huge amount of data generated from genome sequencing projects led to the development of another discipline, called Bioinformatics. Bioinformatics utilizes computer science and mathematics to organize and retrieve the biological data. The biological information such as sequences of nucleic acids and proteins, their structures, post-translational modifications of proteins, etc. are organized and stored in the databases. The databases can be accessed to retrieve the required information for analysis.

The role microscopy plays in understanding biological systems and processes hardly needs any introduction. The first uses of microscopes for observing the biological specimens date back to1660s. It would have not been possible to identify and understand the organization of microorganisms without using microscopy. Light microscopy is used to identify the microorganisms based on their morphology and the specific stains they take up. A routine quantitative application of microscopy is to count the number of different cells per unit volume of blood or any other sample using a hemocytometer. Presence of cells that are not expected in the healthy individuals may be an indicator of anomaly/disease. For example, a simple microscopic analysis of blood sample will identify the sickle cell anemia; presence of pus cells in urine, quantified by microscopy, is an indicator of infection. Light microscopy uses light as the illumination radiation and is perhaps the most familiar form of microscopy. In the simplest microscopic methods, a specimen is illuminated by visible light and observed either against a bright background (bright-field microscopy) or a dark background (dark-field microscopy). Fluorescence microscopy, one of the most commonly used microscopic methods in biological research, has emerged as a very powerful tool for studying molecular processes owing largely to the advancement in optics and discovery of the green fluorescent protein and development of its analogs with different spectral properties. Confocal laser scanning microscopy (CLSM) is a type of fluorescence microscopy that allows imaging of the samples at different focal planes *i.e.* light emitting from below or above the desired focal plane is eliminated. This results in very high lateral resolution and allows determining the spatial localization of the molecules. Total internal reflection fluorescence (TIRF) microscopy is another type of fluorescence microscopy wherein the optics allows imaging of the molecules that are in close proximity to the microscopic slide. The resolution of light microscopes depends on the wavelength of the light used. The smaller the wavelength of the light used, the better the resolution obtained. Wavelength of the visible light imposes a resolution limit of ~0.2 μm on the light microscopes. What it means is that the two point objects closer than ~0.2 μm cannot be resolved used a light microscope.

In electron microscopy, the electrons are accelerated by applying a very high accelerating voltage. The wavelength of the electron beam is inversely proportional to the square root of the accelerating voltage, and wavelengths smaller than 0.5 nm can be generated. This provides around three orders of magnitude improvement inresolution. Scanning electron microscopy (SEM) scans the specimen and provides surface information of the specimen. In transmission electron microscopy (TEM), electrons penetrate into the sample and the transmitted electrons generate the image. TEM, therefore, provides information about the internal structures of the specimen. Both SEM and TEM generally require staining of the specimen with a heavy atom. There have been several advancements in transmission electron microscopy, cryo-electron microscopy (Cryo-EM) is perhaps the most noted one. Cryo-EM allows the imaging of hydrated samples, does not require any staining and can provide resolutions between 5-10 Å making the method useful in studying the structures of biomacromolecules. Advent of scanning probe microscopy, especially the atomic force microscopy, could make it possible doing imaging in solution with resolutions comparable to electron microscope.

References

- "Proof-of-concept study of biomarker development in mice provides a roadmap for a similar approach in humans". www.fredhutch.org. Retrieved 11, February 2020
- De Laet J. (2005). Parsimony and the problem of inapplicables in sequence data. Pp. 81-116 in Albert, V.A. (ed.) Parsimony, phylogeny and genomics. Oxford University Press, ISBN 0-19-856493-7
- Wang, Qing; Raghothama Chaerkady (December 2010). "Mutant proteins as cancer-specific biomarkers.". Proceedings of the National Academy of Sciences of the United States of America. Retrieved 07, May 2020
- U. Grenander and M. I. Miller (2007-02-08). Pattern Theory: From Representation to Inference. Oxford: Oxford University Press. ISBN 9780199297061
- David A. Baum; Stacey D. Smith (2013). Tree Thinking: An Introduction to Phylogenetic Biology. Roberts. p. 442. ISBN 978-1-936221-16-5
- Hogeweg, Paulien (7 March 2011). "The Roots of Bioinformatics in Theoretical Biology". PLOS Computational Biology. 3. 7: e1002021. doi:10.1371/journal.pcbi.1002021. PMC 3068925. PMID 21483479
- Sample, Ian (May 7, 2014). "First life forms to pass on artificial DNA engineered by US scientists". The Guardian. Retrieved 21, March 2020

Permissions

All chapters in this book are published with permission under the Creative Commons Attribution Share Alike License or equivalent. Every chapter published in this book has been scrutinized by our experts. Their significance has been extensively debated. The topics covered herein carry significant information for a comprehensive understanding. They may even be implemented as practical applications or may be referred to as a beginning point for further studies.

We would like to thank the editorial team for lending their expertise to make the book truly unique. They have played a crucial role in the development of this book. Without their invaluable contributions this book wouldn't have been possible. They have made vital efforts to compile up to date information on the varied aspects of this subject to make this book a valuable addition to the collection of many professionals and students.

This book was conceptualized with the vision of imparting up-to-date and integrated information in this field. To ensure the same, a matchless editorial board was set up. Every individual on the board went through rigorous rounds of assessment to prove their worth. After which they invested a large part of their time researching and compiling the most relevant data for our readers.

The editorial board has been involved in producing this book since its inception. They have spent rigorous hours researching and exploring the diverse topics which have resulted in the successful publishing of this book. They have passed on their knowledge of decades through this book. To expedite this challenging task, the publisher supported the team at every step. A small team of assistant editors was also appointed to further simplify the editing procedure and attain best results for the readers.

Apart from the editorial board, the designing team has also invested a significant amount of their time in understanding the subject and creating the most relevant covers. They scrutinized every image to scout for the most suitable representation of the subject and create an appropriate cover for the book.

The publishing team has been an ardent support to the editorial, designing and production team. Their endless efforts to recruit the best for this project, has resulted in the accomplishment of this book. They are a veteran in the field of academics and their pool of knowledge is as vast as their experience in printing. Their expertise and guidance has proved useful at every step. Their uncompromising quality standards have made this book an exceptional effort. Their encouragement from time to time has been an inspiration for everyone.

The publisher and the editorial board hope that this book will prove to be a valuable piece of knowledge for students, practitioners and scholars across the globe.

Index

www.ingramcontent.com/pod-product-compliance
Lightning Source LLC
LaVergne TN
LVHW080734210926
848485LV00003B/557